ISRAEL IRAQ OIL WAR

WINNER OF THE 2003

UNDUE INFLUENCE, DECEPTIONS, AND THE NEOCON ENERGY AGENDA

GARY VOGLER

The
<u>LIBERTARIAN</u>
INSTITUTE

ISRAEL IRAQ OIL WAR

WINNER OF THE 2003

UNDUE INFLUENCE, DECEPTIONS, AND THE NEOCON ENERGY AGENDA

GARY VOGLER

Israel, Winner of the 2003 Iraq Oil War:
Undue Influence, Deceptions, and the Neocon Energy Agenda

© 2024 by Gary Vogler
All rights reserved.

Cover Design © 2024 by Andrew Zehnder

Maps by Joe LeMonnier of MapArtist.com
*Map on back cover depicts the export routes
from Iraq to the Mediterranean Sea since 1934.*

Cover Photo by U.S. Army official photographer
*This photo was taken during the April 2003 mission
to destroy the K3 pump station in Anbar Province,
shutting down the export pipeline to Syria.*

Published in the United States of America by

The Libertarian Institute
612 W. 34th St.
Austin, TX 78705

LibertarianInstitute.org

ISBN-13: 979-8-9896246-2-1

Table of Contents

Foreword by Lawrence B. Wilkerson	i
Preface	v
Disclaimer	ix
Chapter 1: Feedback on *Iraq and the Politics of Oil*	1
Chapter 2: The Oil Agenda – Original Plan	7
Chapter 3: Pipeline Attacks	17
Chapter 4: The Glencore Plan – A New Strategy	25
Chapter 5: Israel's Energy Security – The Motive	33
Chapter 6: The Suspected Mossad Asset	41
Chapter 7: Syrian Oil Production	49
Chapter 8. What Still Haunts Me	55
Appendix A: Unclassified Email – DEPSECDEF Inquiry of Oil for Israel	65
Appendix B: "A Whole New Ballgame Overseas"	67
Endnotes	71
Acknowledgments	79
Also by Gary Vogler	81
The Libertarian Institute	83

Foreword

by Lawrence B. Wilkerson

The Second Iraq War (2003–11) was an American strategic disaster, the worst such failure of the present century and perhaps the worst of the last century as well. In that latter stretch of history, one might ask, "Why not the 1959–75 U.S. venture into Vietnam?" One significant reason is that Vietnam today is a thriving, successful state — despite the American Empire's many depredations. Iraq, on the other hand, is a ruined state now and will be for decades to come — and what most Americans, indeed most Westerners, do not realize is that in many ways Iraq was the most successful state in the Levant *before* the first U.S. invasion in 1990–91 and the even worse, in terms of massive material damage, four-day bombing ordered by Bill Clinton in December 1998. The 2003 invasion put the finishing touches on the almost total devastation of Iraq.

Pointedly, the single aspect of Iraq's prior success that survived this almost total devastation wreaked upon it by the United States and Britain was its oil industry. It's highly likely that its survival was assured because so many different parties and powers — from the arch-criminal Ahmed Chalabi, to the increasingly right-wing government in Israel, to the American Empire itself — were determined to control it. Vivid proof of such determination was provided colorfully by the "mayor of Baghdad," Ambassador Barbara Bodine, who when she entered Baghdad "post" hostilities declared that she could not even find a working telephone, let alone a working ministry — except for the oil ministry.

For that reason as well as many others, the debate over whether oil was a principal reason for the 2003 invasion has waxed and waned, with one camp arguing that it absolutely was, while the other argues the precise opposite. In his first book, *Iraq and the Politics of Oil*, Mr. Vogler goes from the second camp to a very refined version of the first by the book's end. To understand this debate more clearly — and Mr. Vogler's current position within it, now further detailed by this latest book — a look at the U.S.'s long-held strategic approach to the Persian Gulf Region (PGR) is helpful.

First, since President Jimmy Carter publicly declared the U.S.'s strategy in his 1980 State of the Union Address, that approach has never been about controlling the PGR's oil for private companies' profit. Profit such companies — like ExxonMobil and Royal Dutch Shell — certainly have, and it's absolutely the case that the U.S. government promotes such success as part of its global economic and commercial vision; but it does not as a rule wage major wars for it.

Second, what the U.S. does practice diplomacy and, if need be, wage major wars to achieve — and what its strategy for the PGR consists of — is ensuring that (1) the oil flow from the Gulf is consistent, (2) the flow is as near-constant as possible, and (3) the price per barrel of that oil is reasonable. These strategic objectives are not only applicable to the economy of the U.S., but to the economies of our treaty allies, our other close friends, and other members of the global community who do not oppose us in a serious way. Let me repeat: U.S. strategy has never been to seize oil assets for private corporations. So, in this important sense, Mr. Vogler's original position about the invasion not being for oil for private companies — U.S., French, British, or otherwise — was basically sound (somewhat ironically today, the "private company" that has most benefited from Iraq's oil is China).

Mr. Vogler, himself a former soldier, changed his mind in the course of his first book essentially because he discovered that U.S., British, and other soldiers did in fact die and were seriously wounded for oil. And hundreds of thousands of Iraqis died or were wounded, and millions displaced as well, but not for the strategic purpose outlined above. They did so for oil and for Israel.

We need to understand another factor at play in this sordid saga of men and women dying for oil — oil that was not even, directly speaking, for the U.S., our friends, or our allies (minus Jerusalem). That factor is complex, and it centers around the White House.

There is little doubt in expert minds that President George W. Bush was one of the most inexperienced men ever to hold the Oval Office. There is little doubt as well that Vice President Dick Cheney was the precise opposite: an experienced bureaucratic entrepreneur of the highest caliber. There is also little doubt that Secretary of State Colin Powell and Secretary of Defense Donald Rumsfeld — with VP Cheney normally on the latter's team — did not work well together. Moreover, but less well-known, National Security Advisor Condoleezza Rice was out of her element other than constantly working Bush to be his second-term Secretary of State, replacing Powell. In Powell's final call to Bush in early January 2005, Powell unloaded on the president with respect to the utter dysfunctionality of his decision-making team.

Did such a concoction of personalities and tacit alliances lead almost inevitably to malevolent design triumphing over inexperienced presidential leadership in the most truly serious of national security matters, including, most prominently, Iraq? It's difficult to assign Iraq's abysmal state today to anything else, though many have tried to blame all manner of other influences, from ISIS to Iran. Vogler's words in his latest book put a lie to all such obfuscations, half-truths, and official propaganda.

Preface

Why am I writing this book? It has been two decades since the Bush administration's disastrous decision to invade Iraq. Few Americans today defend the decision, except for foreign policy hawks like John Bolton, who was one of its major advocates in 2003. Some prominent Americans believe that it was the biggest foreign policy blunder in the last century. As a participant* in the planning for and execution of the Iraq oil sector for many years, I cannot think of a bigger blunder. And no one has been held accountable.

With 4,489 killed in action (KIAs), 32,223 wounded in action (WIAs), and more than $2 trillion spent in Iraq between 2003 and 2011, our country received zero benefits from these investments. The human cost to the Iraqis was huge as well, at least 250,000 dead. The families of the Americans impacted by that war deserve an honest response to the question "Why did we do it?" This book details the truth about one part of that honest response: the Israeli oil agenda.

My previous book, *Iraq and the Politics of Oil: An Insider's Perspective*, was published in late 2017 by the University Press of Kansas. It describes the U.S. government's involvement in the Iraq oil sector during my five months of prewar oil planning at the Pentagon followed by my 75 months in Iraq under the Department of Defense (DOD). It provides detailed analyses of the successes and failures of the oil sector operations and reconstruction, most of which I personally witnessed. The discussion of an Israeli oil agenda to deliver Iraqi oil to Israel by the end of 2003 was only a small part. For the benefit of any readers who may wish to read *Iraq and the Politics of Oil* as well, this new volume contains footnotes below the relevant passages which identify the related chapters of the 2017 book.

A book review by *Foreign Affairs* in 2018 conveniently overlooked the plan to get Iraqi oil to Israel. This review emphasized to me that I needed to do a better job presenting the evidence of the deceptive oil agenda. I wrote articles in magazines and blogs in an attempt to improve my communication of the deception. These articles are included on the book's Facebook page. In this new book, I take direct aim at the Israeli oil agenda. I may refer to several

---

* I was a Senior Executive Service (SES II) appointee at one point during my Iraq service. I was told that that was equivalent to a 2-star general.

crucial facts on multiple occasions, so that there will be no doubt in my readers' minds as to the context of my argument.

I denied the existence of any oil agenda during my time working for DOD. I was naïve. Many reporters from the mainstream media quizzed me "on background" in early 2003 about possible oil agendas, and I felt that it was my job to convince them otherwise. Ambassador Paul Bremer convinced me to join him and Meghan O'Sullivan in Rachel Maddow's 2014 documentary *Why We Did It*. All three of us took the position that there was no U.S. oil agenda. I believed Rumsfeld and his senior subordinates' claims that there was no oil agenda. As I began researching for my 2017 book, however, I realized that I had been in denial about an oil agenda since 2003.

My first action upon discovering the oil agenda was to contact my old boss, Jay Garner, the head of the Office of Reconstruction and Humanitarian Assistance (ORHA). I apologized to Jay for having failed to advise him accurately about an oil agenda back in 2003 when we could have done something about it and probably saved lives. Jay told me to stop beating myself up. He commented that we were in execution mode back then, getting a hundred tasks a day and only accomplishing a portion of them. We did not have time to study from whom and why we were getting various tasks. He mentioned that everyone had an agenda for that war, and it only made sense that the neocons had an agenda to get oil to Israel. I cannot stop beating myself up. That war continues to haunt me.

The last chapter of my 2017 book, titled "What Haunts Me," summarized what I learned about the oil agenda in my research. In hindsight, while I did introduce the matter of an oil agenda, I did not do the subject justice. Because there is so much more to the oil agenda that has not been reported, it is my goal to continue investigating so that I can add critical information to what has already been written.

Is it a matter of coincidence that Israel and China, the two biggest winners of the Iraq oil sector since 2003, are also the two countries that lobby our D.C. lawmakers the most? Discussions I have had with Iraq oil reporters and others with insight into Iraq's oil sector think that it might be a coincidence in the case of China, since the rules used by Iraq during its 2009–10 oil field auctions favored Chinese oil interests. So, while I continue to have concerns about China, it is for others to pursue those coincidences. China is not the subject of my book. The detailed case I make regarding Israel is not based on coincidence, and it is the Israel case I will pursue.

Our country — especially our military — invested huge amounts of blood and treasure during the Iraq War, and our country received zero benefits. Those of us who risked our lives were naïve about the reasons for that war and very easily could have been part of those investment statistics. It is time to share more of what I have learned about the oil agenda, who benefited, and why.

Disclaimer

Like many Americans, I have always been and remain pro-Israel, just as I am pro-U.K., pro-Canada, pro-Australia, and pro- any other countries with whom we share many common national interests. Israel has been a longtime friend of the United States in a very difficult part of the world. I have always respected its military, government leaders, and lobbyists.

In my four years of working in Saudi Arabia for Mobil Oil in the late 1980s, I learned that other than buying their oil, Americans share very few national interests with the Saudis. My Army Reserve unit during that time was at the European Command (EUCOM) headquarters in Stuttgart, Germany, where I spent two weeks training with the EUCOM staff every year between 1988 and 1995. My assignments involved working with allied military officers, and Israel was under the EUCOM portfolio at the time. I gained a healthy respect and value for all our allies, including the Israelis.

All my criticism in this book is directed at the Bush administration for its failure to vet the conflicted neoconservatives working on behalf of Israel inside the U.S. government. The Bush administration failed me, and I have seen no effort to enforce or fix the laws that enable abuses by Israel's powerful lobby in Washington.

Chapter 1: Feedback on *Iraq and the Politics of Oil*

*Absent the neocons, the Iraq War
would not have happened.*

It is important that my story be told. It is the history of how neoconservatives inside the Bush administration worked secretly with the Israel lobby and the Israeli government to fix Israel's energy security crisis. Israel has significantly benefited since the 2003 Iraq War, while our country has incurred significant costs. My role was that of an unwitting contributor to their success. The events, assessments, and conclusions described in this book are from my personal experience.

Authors typically get thoughtful feedback from their readers, and sometimes readers may have profound insight regarding the subject matter. My 2017 book on the Iraq oil sector provided me with many such excellent observations from some very prominent readers. I am writing this new book aided by the thought-provoking comments and encouragement that I have received from those readers. Here are three examples of such feedback.

Retired Senior Vietnam-Era General Officer

A retired senior general officer who mentored some of the most successful generals of my generation, including John Abizaid and David Petraeus, wrote me an email in the spring of 2018. He commented that while he enjoyed the book, he had to read it twice because he overlooked much of the detail during his first read.

He remarked that I had been naïve about Washington special interests during my five months of prewar oil planning in the Office of the Secretary of Defense (OSD) and during my 75 months of service in Iraq. I agreed. He told me not to take offense at this observation. He shared that he had also been naïve during his entire 38 years on active duty, as had every general officer he knew.

The fact that we have so many naïve senior officers in the military did not surprise me, and it should not surprise anyone. What pleasantly surprised me was that this highly respected retired senior general officer was so willing to admit it. I have always respected honesty. I know many retired flag-ranked military officers, and I have often wondered how many are comfortable enough with their egos to make similar admissions and do something about it.

Senior Civilian Political Appointee

The next comment came from a senior civilian political appointee at the Pentagon. I became acquainted with him during my first tour in Iraq while working for the Coalition Provisional Authority (CPA) in 2003–04.

A few of the senior Republican political appointees with whom I had worked in Iraq refused to return my emails once my book was published, so I was surprised to see that this person reached out in a phone call. I immediately asked him what made him so different from the others who shunned me. He responded that while their loyalties were to the Bush dynasty, his loyalties were to the country and to the Republican Party.

He proceeded to commend my courage for writing the truth about the oil agenda, but then he added that doing so destroyed any chances that I may have had for a career in government. I asked him whether this was because some powerful people in Washington did not believe my story of Iraqi oil being delivered to Israel. No, he replied, these political folks knew that my conclusions were accurate, but they would not want that information to find its way into the public discourse. He said that it was deemed by some very powerful interest groups to be politically sensitive.

Then he lowered his voice to say that he wanted to share something with me, but that he would deny it if I ever quoted him by name. He alleged that Donald Rumsfeld was quoted by one of his close advisors as stating during a moment of frustration that *he* did not run the Pentagon, but AIPAC (the American Israeli Public Affairs Committee) did. Think about that comment. The Secretary of Defense himself was stating that a powerful foreign lobby was running the Pentagon.

West Point Classmate

In September 2018, I attended my 45th West Point class reunion at the Academy. The weather was perfect that weekend, and several classmates attended the festivities. We attended a football game, a parade by the cadets, and meetings with the superintendent. Many of my classmates have spent their careers in the military for several decades and have served at the highest levels of the Army.

One such classmate completed 26 years as an infantry officer and retired as a colonel. He approached me during one of our social events, saying that he had recently read my book and found it interesting. He had one important question that he wanted to ask: whether I would have volunteered as a civilian for those six-plus years in Iraq if I had known in 2003 what I knew today.

I paused, then looked him in the eye, and said, "Not only 'No,' but 'Hell no!' Why would I risk my life for 75 months in a war zone on the receiving end of hostile fire — only to supply cheap gas at the pump for Israelis? Or to fill the bank accounts of those conflicted neocons both inside and closely linked to the Bush administration? The one neocon on our prewar planning team even refused to go to Iraq. He was the only person on our team who refused to volunteer."

The retired infantry colonel said that he certainly understood my feelings. Then we both lamented that those in uniform do not have a choice. Like many civilians working for the U.S. government, I could have resigned at any time during my tours and immediately boarded the next plane out of Iraq. Those in uniform did not have that choice. They were there for the duration of their mission.

The 4,489 KIAs and the 32,223 WIAs did not have a choice. They were so trusting about why we went to war and had assumed that our political leaders were motivated just as we were, for God and country. Many made the ultimate sacrifice, and for what? Again, our country has realized zero benefits from that war. In fact, the consequences continue to surface years later, particularly in the form of post-traumatic stress disorder (PTSD) afflicting many veterans.

Bush's Motivations for War

So, why was President Bush so highly motivated to go to war with Iraq? There was no doubt among the prewar planning team that the president was going to war, no matter what the United Nations Security Council (UNSC) decided. Without UNSC concurrence, most Americans would feel that a war would be considered illegal, and our country claims to follow the rule of law. But the UNSC's refusal to agree to the invasion seemed to be the only obstacle.

According to the U.K. whistleblower Katherine Gun, the U.S. government was desperate for a vote from the UNSC in 2003. Gun was prosecuted for revealing classified information to the U.K. press that explained how the U.S. wanted any dirt on the members of the UN Security Council in order to pressure them for a vote in their favor. A notable quote from Gun was: "I work for the British people. I do not gather intelligence so the government can lie to the British people."[1] The film *Official Secrets* (2019) gives an overview of just how the U.S. sought the information from the U.K. government.

I wrote in my 2017 book that, short of reading Bush's mind, we would never really know why he was so highly motivated. My conclusion has since changed. I am now convinced that President Bush had many reasons for war, but that there were two overarching reasons. The first was to avenge the (alleged) assassination attempt on his father in Kuwait in the early 1990s. The second was to ensure that the Israel lobby did not prevent him from winning a second term. Bush was convinced that his father lost to Bill Clinton in 1992 because he had lost the support of the Israel lobby. Bush 41 had received 27% of the Jewish vote in 1988 but only 13% in 1992.[2] Bush 43 was not going to repeat the mistakes of Bush 41.

Two CENTCOM senior military officers whom I interviewed in 2016 told me that Bush, when advising Middle East leaders prior to the war, remarked that he would avenge the alleged assassination attempt on his father. The conversations between Bush and Middle East leaders were reported and briefed to the senior CENTCOM staff.

Former CIA analyst and Saddam expert John Nixon, in his book *Debriefing the President: The Interrogation of Saddam Hussein* (2016), also stated that Bush, whom he had briefed many times before the war, wanted to avenge the alleged assassination attempt on his father.[3] I met with Nixon after my book was published in order to understand his theories regarding the president's motivation. By citing a multitude of sources during our two-hour lunch, Nixon had me convinced of the revenge narrative.[4]

Robert Draper, in his book *To Start a War: How the Bush Administration Took America into Iraq* (2020), claimed that even Jeb Bush stated in 2006 that all the destruction witnessed during his tour of Iraq was justified because Saddam had allegedly tried to assassinate his father in Kuwait in the early 1990s.[5] (Investigative reporter Seymour Hersh cast major doubt on the assassination plot narrative in a piece he wrote in late 1993.[6])

The second overarching reason for war — keeping the Israel lobby happy — was satisfied by staffing key positions throughout the government with

AIPAC-backed appointees. Many of these neoconservatives held important positions at the Pentagon, on the National Security Council (NSC) staff, and in the Office of the Vice President. They were the administration's most enthusiastic drivers for an Iraq War.

Tom Friedman, the Pulitzer Prize-winning *New York Times* journalist, had a lot of credibility among the generals and foreign service officers with whom I worked in Iraq. A couple of weeks after the war started, Friedman placed the primary blame for the Iraq War on these neoconservatives who pushed for war.* Reporters from the Washington, D.C., office of *Haaretz* interviewed Friedman in early April 2003. Friedman understood then just what it would take many of us several years to comprehend fully — namely that, absent the neocons, the Iraq War would not have happened:

> It's the war the neoconservatives wanted. It's the war the neoconservatives marketed. Those people had an idea to sell when September 11 came, and they sold it. Oh boy, did they sell it. I could give you the names of twenty-five people who, if you had exiled them to a deserted island a year and a half ago, the Iraq War would not have happened.[7]

Friedman did not name the neoconservatives, but many were occupying important positions in the Bush administration, including Paul Wolfowitz, Doug Feith, Lewis Libby, Richard Perle, Elliott Abrams, Michael Ledeen, David Wurmser, and David Frum, among others. (Richard Perle, Doug Feith, and David Wurmser had led the 1996 "Clean Break" strategy report, completed for Israeli leaders, which called for regime overthrow in Iraq, Syria, and Iran.[8]) During a conference at the University of Virginia in 2002, the year before our invasion, State Department Counselor Philip Zelikow also revealed that Israel's security problems were an important factor for a push for war.[9]

Why were the neoconservatives so motivated to go to war in Iraq? Almost all were ideologues committed to Israel.[10] The one that I was most familiar with, Michael Makovsky, convinced me that he was, in fact, an agent for Israel, and I discuss my reasons later in this book.

Israel had entered into a severe energy security crisis in the late 1990s, which the 2003 Iraq War solved. While the U.S. recognized zero benefits from the Iraq War, our Middle East friend has experienced quite a windfall.

---

* Also recounted in Chapter 20 of *Iraq and the Politics of Oil*.

This book discusses the neocon oil agenda for getting Iraqi oil into Israel. It discusses the who, what, where, why, and how of that agenda. It covers how the initial plan of opening a pipeline between Kirkuk, Iraq, and Haifa, Israel, would never work, so an alternative plan was executed. This book discusses how the U.S. continues to support operations in Syria that provide more oil to Israel. It covers the history of Lewis Libby's role in Israel's energy security, first as Marc Rich's lawyer and later from his position in the Office of the Vice President. It also covers how a suspected Mossad asset, Michael Makovsky, was placed in an important position at the Pentagon and given a top-secret SCI clearance to help dictate the oil agenda. Today, he is the CEO of the Jewish Institute for National Security of America (formerly, "Affairs") (JINSA), located a few blocks from the White House. Several years ago, I spoke with an intelligence resource who described that organization as a front for Mossad. The JINSA website states that their mission is to fund trips to Israel by retired U.S. general officers to meet with Mossad.[11]

I have always considered myself pro-Israel, and I still do today. However, having witnessed and experienced many abuses by the Israel lobby in matters surrounding Israel's energy security, I am convinced that greater transparency of foreign lobbyists is necessary on the part of the U.S. government. New laws need to be enacted to protect patriotic citizens from these grossly conflicted government appointees who represent foreign interests. Furthermore, existing laws, such as the 1938 Foreign Agents Registration Act (FARA), need to be enforced when dealing with Israel. As you read the following chapters, you will see the various instances where a very powerful, supposedly "friendly" foreign lobby abused the trust of U.S. citizens in order to pursue policies for their own country at a great cost to ours. The Bush administration enabled this abuse of power. No one has been held accountable, and no laws have since been enacted to prevent any such abuses of power in the future.

Chapter 2:
The Oil Agenda – Original Plan

*Urgency to get Iraqi oil to Israel
by the end of 2003.*

A plan of simply opening the valves on a pipeline in the Iraqi desert — after the pipeline had been closed for more than 50 years — may have seemed foolish to most people, but the Israeli government and the neoconservatives in the Bush administration intended to make it happen. In June 2003, Benjamin Netanyahu, the Israeli finance minister at the time, visited London to find private investors to whom he could pitch the repairing and expanding of what he described as "not a pipe dream."[1]

Ever since the Baba Gurgur oil field was discovered just north of Kirkuk in October 1927, oil has played a role in Iraq's economic history.* In 1934, Iraq began exporting oil to the Mediterranean through a pipeline to Haifa and also to Tripoli, Lebanon. At the time, Haifa was under British control, and Tripoli was under French control. Deliveries to Haifa stopped in 1948, when Iraq ordered the Iraq Petroleum Company (IPC) to stop delivering to the new state of Israel after the first Arab-Israeli War.[2]

IPC had a monopoly concession for all of Iraq's oil production. It produced and transported the oil to its customers in the Mediterranean and Europe while paying small fees to the Iraqi government. When the shipments to Haifa stopped, IPC needed a new export channel for handling its production from northern Iraq.[3] IPC constructed a new pipeline to Baniyas, Syria, in 1952 in order to get that oil to the Mediterranean.[4] This pipeline was operational until April 2003, with one exception: it was damaged during the First Gulf War of 1991 (U.S. Operation Desert Shield/Desert Storm) when the U.S. bombed the K3 oil facility, known as the "K3 pump station," in Haditha. The Iraqis repaired the pumps and continued to export oil through Syria from the late 1990s until the 2003 invasion.

In March 2003, Deputy Secretary of Defense Paul Wolfowitz ordered General Franks to destroy the K3 pump station. His order conflicted with

---

* This is covered in detail in Chapter 4 of *Iraq and the Politics of Oil.*

the then-recently declassified written order by President Bush's cabinet that no oil infrastructure should be intentionally attacked during the invasion. The Syria pipeline was specifically identified *not* to be attacked.

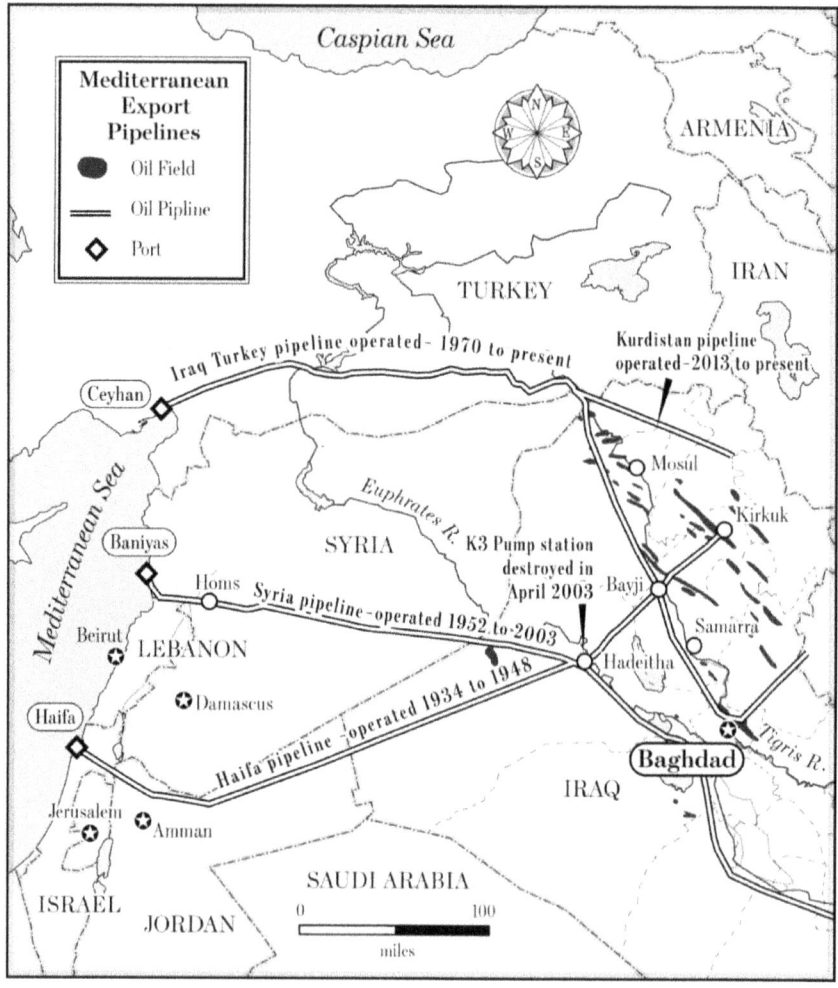

Figure A: Export pipelines from Iraq to the Mediterranean Sea. (MapArtist.com)

Reopening the old pipeline from Kirkuk to Haifa was the original plan sold to the neoconservatives and the Israeli government in the late 1990s by Ahmed Chalabi, the leader of the Iraqi National Congress (INC). I first read about this Chalabi/neocon plan in 2008, in the book *Losing Iraq: Inside the Postwar Reconstruction Fiasco* (2006) by David Phillips.[5] Chalabi delivered speeches to various Israel lobby organizations in Washington, D.C., most notably to the Jewish Institute for National Security Affairs (JINSA) in June 1997, at a time when Israel was still paying a large premium on oil imports. Chalabi promised to reopen the pipeline to Haifa, so long as the lobby could get the U.S. military to remove Saddam and appoint *him* as prime minister.

Israel had previously enjoyed an oil supply arrangement for several decades with Iran. Iran would supply its Mediterranean customers with its own oil, shipped around the Arabian Peninsula to the Red Sea, and then to the Gulf of Aqaba and the Israeli oil port of Eilat. From Eilat, the oil was transported through Israel in a very large but secretive pipeline. The Eilat-Ashkelon pipeline could carry more than a million barrels a day. The Israeli port of Ashkelon loaded tankers destined for Mediterranean countries. Israel retained a portion, something less than 250,000 barrels a day, for its own use.[6] This supply agreement, which provided Israel with the oil it needed at a discounted price, ended in 1995, and the details of why it ended will be covered later in this book. Israel began to source its oil from elsewhere, paying roughly a 25% premium. Chalabi and the neoconservatives recognized that Iraqi oil piped from Kirkuk to Haifa would be a huge relief for Israel.

After reading Phillips's chapter about Chalabi, I immediately contacted an engineer who had worked with me in Baghdad. She was an Iraqi-British citizen who had worked in the INC for Chalabi in the 1990s. I asked whether Chalabi had made such a promise to open this pipeline that had been shut down for more than 50 years. She replied that, yes, since Chalabi knew what the Israel lobby wanted to hear, he told them so. She further mentioned that Chalabi was easy to believe and that they wanted to trust him.

My skepticism remained. I was not convinced that such a plan was believable or doable. Rumsfeld, Wolfowitz, and Feith had all convinced me that there was no oil agenda. Why would they mislead me? I believed them. Their deception was convincing. I was in denial! So, what finally changed my mind?

The research for my 2017 book uncovered several articles and interviews that were previously unknown to me. I subscribed to the Israeli newspaper *Haaretz* so that I could search for older, archived articles. A few *Haaretz* articles particularly caught my attention. The first was the previously mentioned Tom Freidman interview from early April 2003.[7] Even though Friedman was convinced that the neoconservatives were the most critical pushers for war both inside and outside of the Bush administration, he did not discuss their motive. Just why were they such strong advocates for war?

Let me share some of the new facts I learned during my research for the 2017 book that helped me complete the oil puzzle. When I compared these newfound facts to my personal experiences at the Pentagon and in Iraq, I could no longer deny an oil agenda. It was not an easy change for me. The more I learned about the agenda, the angrier I became. I could not sleep at night. My blood pressure boiled. My wife rushed me to the hospital one Sunday afternoon in 2015. We thought that I was having a stroke.

A *Haaretz* interview with Israeli Infrastructure Minister Joseph Paritzky took place on March 31, 2003. While our own troops were still fighting some intense battles with the Iraqi army south of Baghdad, Paritzky was publicly bragging about the huge windfall that Israel would experience from the war. He identified the agenda, the players involved, the timing, and, most importantly, the hidden motive: getting Iraqi oil to Israel by the end of 2003 to relieve the severe 25% premium.[8] The U.S. invasion and a reopened Kirkuk-Haifa pipeline would remedy this problem. Paritzky said that senior civilians at the Pentagon were communicating with him on this objective.

Paritzky was not the only member of Israel's cabinet in 2003 publicly making statements about getting Iraqi oil. Finance Minister Netanyahu was in London in June promoting the Haifa pipeline to potential investors. "It won't be long when you will see Iraqi oil flowing to Haifa. It is not a pipe dream," Netanyahu told the investor group.[9]

One article, "How Ahmed Chalabi Conned the Neocons" by John Dizard, quotes Marc Zell, Doug Feith's former law partner who was still trying to develop his Israeli law practice based on Chalabi's promises of ending Iraq's boycott of trade with Israel and of a rebuilt pipeline. Quoting Marc Zell: "[Chalabi] said he would end Iraq's boycott of trade with Israel and would allow Israeli companies to do business there. He said the new Iraqi government would agree to rebuild the pipeline from Mosul in the northern Iraq oil fields to Haifa." But Chalabi, whom Zell calls a "treacherous, spineless turncoat," delivered on none of these promises.[10]

Several experiences during my prewar planning and time in Iraq did not make any sense to me at the time, but made perfect sense once I was finally able to complete the oil agenda puzzle.

Enter Makovsky

Michael Makovsky was assigned to our prewar planning team* in October 2002. We had five key members of the team, plus a supervisor. Makovsky had no relevant prior oil or government experience, but everyone else on the team was highly qualified to be there. Clarke Turner, for instance, was an experienced Army officer from the 1991 Gulf War. He had previously been working as a civilian for the Department of Energy on oil production in Wyoming for several years. Clarke and I understood everyone's role on the team except for Makovsky's, so in the first week after our team was activated, we went to Mike's cubicle to ask him about his experience in the oil business. Makovsky only responded that he was on the team because he was Doug Feith's brother-in-law. He then ignored us and continued looking at his computer. As we walked away, I asked Clarke if there were any government nepotism regulations prohibiting hiring a brother-in-law. Clarke smiled and said that Mike was not related to Feith; he just did not want to share his credentials. Makovsky was on the team because of *whom* he knew and not *what* he knew.

During our prewar planning, Makovsky organized several meetings with private sector companies. All these private sector folks seemed to know Richard Perle and asked Makovsky about him. After one of the meetings, I asked who this Richard Perle person was. Makovsky only said that he was their "godfather." Perle was also known to others as the "Prince of Darkness."[11] He had been appointed as the chairman of the Pentagon's Defense Policy Board in 2001 and helped to place many Zionist ideologues throughout the Pentagon, such as Deputy Secretary of Defense Paul Wolfowitz.[12] Wolfowitz, one of the primary architects of the 2003 Iraq War, had earned the title "Wolfowitz of Arabia."[13] Another Perle appointee, Under Secretary of Defense for Policy Douglas Feith, whom CENTCOM commander General Tommy Franks described as "the dumbest fucking guy on the planet,"[14] had worked for Perle during the Reagan administration. In 1996, he co-authored the strategy paper "A Clean Break" for Israeli Prime Minister Benjamin Netanyahu.[15]

Our team eventually learned that Makovsky had briefly worked for a small oil trading company while attending graduate school. We also learned that he was very bright and an excellent writer, but we recognized that he marched

---

* The team is described in more detail in Chapter 2 of *Iraq and the Politics of Oil*.

to the beat of a different drum from the rest of us. He seemed to be looking out for Israel's interests. In fact, he often stated during meetings that the national interests of Israel and the U.S. were identical. None of us believed that statement, but he must have, either because he stated it so often or because he wanted to convert us to this view. How many Israeli-American dual citizens working inside of the U.S. government believe this statement?

One example of Makovsky's looking out for Israel's interests occurred in December 2002, when he called a meeting of the group to discuss Iraq's export pipeline through Syria. We had already agreed to the U.S. policy that no oil sector infrastructure would be intentionally attacked during the warfighting, but Makovsky proposed that there be one exception to that policy. He wanted this export pipeline destroyed. His reason was that the Syrians had helped Saddam smuggle oil outside of the UN's oil-for-food program. He conveniently neglected to mention that Jordan and Turkey had also helped Saddam smuggle oil.

We advised him that his reasoning was faulty. The pipeline was owned by Iraq and used for Iraq's benefit. Syria was only paid a small transit fee for pumping oil to the port of Baniyas. Less than 5% of the punishment would be against Syria, while 95% would be against the future government of Iraq. We tried to placate Makovsky's demands by proposing that, instead of destroying the pipeline, we ought to leverage any future exports to get Syria's cooperation. This policy was approved by the president's cabinet in late 2002 and included in Under Secretary Feith's January 16, 2003, report to Secretary Rumsfeld (see Figure B). The full document, declassified on May 19, 2015, is included in my Amazon Kindle book, *Lessons Learned – The Iraq Energy Sector* (2016).[16]

The Oil Agenda – Original Plan

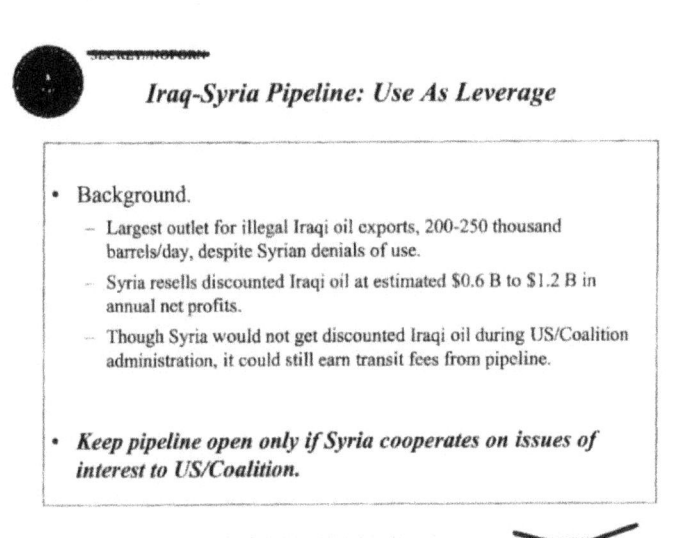

Figure B: Excerpt from Under Secretary Doug Feith's January 16, 2003, report to Secretary Rumsfeld.

April 2003 K3 Pump Station Attack

The cabinet's written policy should have ended further discussion of this issue, but it did not. Makovsky approached me a few days before my deployment to Kuwait in March 2003. He said that I needed to attend a video teleconference in less than an hour in the Pentagon's conference room. It was during this meeting with General Franks that Wolfowitz countermanded the cabinet's written policy and ordered Franks to destroy the Syrian export pipeline.

It was a mission that was not necessary for the U.S. forces, but U.S. soldiers died supporting it. It was a mission that has since cost Iraq billions of dollars in lost export revenues. It was a mission that hurt our credibility with the Iraqi oil ministry because we refused even to consider repairing the $50 million in damages to the pipeline pump station.[17] In my opinion, this mission was executed strictly to hurt Israel's enemy, Syria, and to help solidify Israel's position if the Kirkuk-Haifa pipeline were to become active again.

The saddest part was the loss of life. The 75th Rangers were assigned the mission to destroy the K3 pump station,* which was located very close to the Haditha Dam in Anbar Province where the 3rd Battalion, 75th Rangers had set up their base camp. On April 3, three soldiers were killed from wounds sustained during a suicide bomber attack. The three soldiers were Captain Russell Rippetoe, Specialist Ryan Long, and Staff Sergeant Nino Livaudais of 3rd Battalion, 75th Ranger Regiment.[18]

Makovsky's Phone Calls

Makovsky phoned me during the first week of July 2003 at my office in Baghdad.† He stated that senior civilian leaders at the Pentagon needed to know the status of the pipeline to Haifa. I did not recognize the initial sensitivity of his request because he disguised the name Haifa by using another pronunciation, calling it *Hafu*, which I assumed was somewhere in Iraq. When I asked the oil ministry CEO, he immediately recognized the name and warned me as to the sensitivity of asking about the status of this pipeline. Making such an inquiry about this old pipeline, he said, would only confirm many Iraqis' suspicions that the only reason the U.S. had invaded Iraq was to take their oil and send it to Israel. I told the CEO to ignore the request unless I got back to him. When Makovsky phoned me back later that day seeking the pipeline's status, I requested that he put his request in a formal email that I would review with Ambassador Bremer before seeking the information. Makovsky hung up the phone!

Several weeks later, I learned that Makovsky had bypassed me in a phone call directly to the oil ministry CEO. The CEO had determined that the pipeline no longer existed, as much of the pipe in the ground from 1948 had since been excavated and used elsewhere in Anbar Province.

July 2003 Pipeline Attacks

Later during the month of July 2003, two major pipelines supplying crude oil from the northern oil fields to the Baghdad refinery were attacked. The attacks shut down the refinery for over a week, causing significant shortages of gasoline and diesel product for Iraq. These attacks were significant in that they demonstrated how easy it was for an insurgency to create instability

---

* The attack is described in more detail in Chapter 20 of *Iraq and the Politics of Oil*.
† Makovsky's call is described in more detail in Chapter 10 of *Iraq and the Politics of Oil*.

within Iraq. The gas lines throughout Baghdad were several miles long. Frank Miller of the National Security Council ranted in a video teleconference about how the photos of gas lines were front page news in the United States, broadcasting how the Bush administration had lost control of Iraq.

In 2003, we did not understand why the Iraqis were attacking their own pipelines, but two years later, I learned that the motivation for that first attack on the pipelines was that some workers at the Baghdad refinery were convinced that the United States was opening a pipeline to Haifa to send Iraqi oil to Israel. These Iraqi workers were willing to attack their own oil infrastructure in order to prevent oil from going to Israel, a country they have boycotted since 1948.

Lewis Libby and Ahmed Chalabi in Constant Contact

Lewis Libby also played a role.* Our prewar planning team prepared several policy presentations that were first reviewed by Doug Feith. Feith would review the draft presentations before we sent them for presidential cabinet-level review. Sometimes Feith annotated in the margins that we should review certain policies with "Scooter" before progressing in the approval process. I initially thought that Scooter was another classified Pentagon prewar team like our team, but I quickly learned that Scooter was Irve Lewis "Scooter" Libby, who was chief of staff to Vice President Dick Cheney.

Why did we have to review Iraq oil policies with the vice president's chief of staff? Was it because of Libby's history of working so closely with the notoriously corrupt Marc Rich, the supplier of most of Israel's oil for over two decades? Or was it because Libby was going to be the communications link to Chalabi once we arrived in Baghdad? I suspect that it was both.

Lewis Libby was the law partner of Richard Nixon's personal attorney, Leonard Garment. Under Garment's direction, off and on over a period of 16 years, Libby was Rich's attorney.[19] Libby made a small fortune as the lawyer of Rich back in the 1980s and 1990s. Rich, known within the oil trading business as one of the most successful Middle East oil traders, made billions of dollars from a secret pipeline through Israel that transported Iranian oil to the Mediterranean. Rich transported this Iranian oil while Iran held American hostages and during the following years when Iran was under

---

* For more on the relationship between Chalabi and the Office of the Vice President, see Chapter 10 of *Iraq and the Politics of Oil*.

U.S. sanctions. Such transactions after the Iranian Revolution would have been illegal for a U.S. company, but although Rich was a U.S. citizen, his company was registered in Switzerland.[20] In 1983, Rich was indicted by a grand jury for tax evasion, wire fraud, racketeering, and oil deals with Iran, but he fled the United States after the indictments.[21]

Lewis Libby had a reputation of being close to the Israeli government. According to Jack Straw, who was the British foreign secretary from 2001 to 2006: "It's a toss-up whether Libby is working for the Israelis or the Americans on any given day."[22] Can you imagine a senior foreign diplomat stating that about *any* person in our government, let alone the vice president's chief of staff?

Secure encrypted communications were installed between Ahmed Chalabi's offices in Baghdad and the Office of the Vice President, where Lewis Libby was chief of staff. Even though a STU-III (secure telephone unit, third generation) looks like an old telephone from the 1980s, it is a highly technical device that enables encrypted secure voice transmission. One was installed in Chalabi's Iraqi National Congress headquarters for communications to the vice president's office in Washington. An Army colonel was sent to support Chalabi with these secure communications. Unfortunately, that colonel had to be evacuated from Iraq on emergency leave. Rather than sending a new officer to support this effort, CENTCOM sent a colonel to retrieve the STU-III. Chalabi's headquarters did not want to give up the instrument, so the process took several weeks.

The Office of the Vice President falls under the legislative branch of government, according to Vice President Dick Cheney's lawyer, David Addington.[23] This is relevant to my story because the legislative branch is exempt from the records retention laws that govern the executive branch, and one cannot request documents from the legislative branch under the Freedom of Information Act (FOIA). Therefore, any secret or unofficial agendas could be easily accomplished in a clandestine manner without fear that a future FOIA action might make them public information. Logs of the unofficial oil agenda discussions during secure telephone communications between Ahmed Chalabi's office in Iraq and the Office of the Vice President will never be made public, and they were never shared with those of us in Iraq executing the official oil policy approved by the president's cabinet.

Chapter 3: Pipeline Attacks

*"If the Americans were sending our oil to Israel,
I would help you blow up our pipelines."*

The insurgency officially started in the oil sector during the third week of July 2003. My phone rang late one night. It was Thamir Ghadhban, the CEO of the Iraqi oil ministry: "Gary, two of our crude oil pipelines have been attacked within a two-hour period. This is different from all the previous attacks. The others were primarily for economic reasons, to steal the oil. This attack is intended to shut down the Doura refinery. It cuts off all supply of crude oil to Doura."

This was significant. Iraq had just recovered from the severe shortages of petroleum products in May and June. Demand was now being met by refinery production, but there was no inventory build, no stocks of gasoline. A shutdown of more than a day would translate into shortages, long lines at the gas stations, and security instability within the population. This always resulted in more attacks on the U.S. forces.

The "Good News Only" Meeting

The 8 a.m. meeting the next morning was my first opportunity to share the significance of this event. Ambassador Bremer was the senior civilian leader of the Coalition Provisional Authority (CPA) in Iraq, and he had a meeting every morning of all his direct reports and many others. I spoke up: "Sir, we had an attack on the two crude oil pipelines feeding the Doura refinery last night. The attacks were at two different locations but occurred at about the same time. This will shut down the refinery for some period, and shortages of products will be noticeable almost immediately. The sector needs help. We do not have the resources to secure these pipelines."

Ambassador Bremer looked grim. He did not appreciate bad news brought up during this 8 a.m. meeting of about 50 people. I was later informed that this meeting was known as the "good news only" meeting. I had violated Bremer's unwritten rule, but it was my first opportunity to report bad news that would impact everyone in the meeting within a day or two. Bremer asked Major General Carl Strock to get with Brigadier General

Dan Hahn and me immediately after the meeting and to report back to him later.

Strock, Hahn, and I stayed after the Bremer meeting, and I told them everything I knew. The facts indicated that it may have been an inside job, as only a very limited number of current or former employees of the oil ministry knew the routing of these pipelines. All maps had been controlled documents under the previous regime, and they still were.

I would learn a couple years later that it was, in fact, an inside job. An oil ministry official who visited the U.S. in 2005 shared that shortly after the incident, when he met with the only employees who knew the routing of these pipelines, one of them admitted to attacking the pipelines. His motivation was that the Americans were sending Iraqi oil to Israel, and he produced an Arabic newspaper article that was quoting from an international newspaper. "What was your response when you heard this?" I asked. The Iraqi official stated, "I chastised them. I said, 'You idiots, the Americans are not sending our oil to Israel. If they were, I would help you blow up our pipelines.'"[1]

After the meeting with Generals Strock and Hahn, I immediately left to meet with the oil ministry CEO, Thamir Ghadhban. He was still getting updates from his managers on the extent of the damage. It looked very bad.

Resistance to Importing Products from Syria

Ghadhban and I discussed our options. The U.S. forces had been notified and were issuing task orders to contractors for buying and importing petroleum products as they had done during the May crisis. The CPA was convinced that the shortages were fixed once the domestic refiners were back in operation. Unfortunately, this latest attack changed our strategy. It would take a week or so to mobilize sizeable volumes and get the needed petroleum products into Baghdad. What were our other options?

The CEO had a good idea: "Gary, your contractors can import from Turkey, Jordan, Saudi Arabia, and Kuwait. That is good. We have another neighbor to the west who might be able to help with this problem. SOMO [the State Oil Marketing Organization] has a history of working with Syria. How about SOMO take the lead on importing products from Syria?" Iraq's SOMO, which is responsible for all exporting and importing of oil, had developed excellent contacts in Syria, as well as in Turkey and Jordan, while executing their smuggling operations before 2003 to bypass the sanctions regime.

I knew that there would be resistance to buying products from Syria. This resistance would come from inside the U.S. government and most certainly from the senior civilian leadership at the Pentagon. I knew that the neocons' war agenda included taking Syria and Iran after we took Iraq.[2] They would resist any engagement with the Syrian oil ministry, even if it could save the lives of U.S. soldiers. With that in mind, I instructed the CEO accordingly.

"We need the product, and we need it quickly. Have SOMO check with their contacts to get an understanding of what products and how much they can supply. Do not make any commitments. This is sensitive, and any decision to proceed will require approval from Ambassador Bremer," I said. After this meeting with the CEO, I returned to the CPA headquarters back in the Green Zone.

We scheduled the weekly conference call that evening to update my boss, Phil Carroll. Phil was at his home in London, taking a break from Iraq. There were a few others on the call outside of Iraq. Two representatives on the National Security Council were facilitating the call from the White House. Michael Makovsky at the Pentagon had also joined the last couple of conference calls. He reported to Doug Feith and Paul Wolfowitz, keeping them informed about the status of the oil sector.

Phil welcomed everyone to the call and quickly asked CEO Ghadhban to provide an update on the status of shortages and the current plan to fix the problems. Ghadhban went into the details on locations of product volumes, distribution to specific areas with acute shortages, import supplies committed by the U.S., and an estimate on the return of production of the Doura refinery. As an afterthought, he said, "Gary, SOMO contacted the Syrians today about product sourcing and…" At that point, Makovsky interrupted by shouting over the phone that the CEO could not talk to the Syrians.

There was a long pause of silence when Makovsky stopped his tirade. Then, Phil spoke up. "Well, that's all for this evening. We'll talk again next week." I hung up my phone and sat there in disbelief of the arrogance of Makovsky.

Not more than ten seconds passed before my phone rang. It was Phil. "Gary, that was the last conference call for this group, no more. The official reason is that the calls are no longer needed. The real reason is that I will not tolerate a low-level staffer, sitting in a safe and secure refuge, insulting a respected Iraqi oil executive like Thamir Ghadhban, especially as we sit here on the brink of fuel shortages and a widespread panic."

The Neocons Failed to Stop the Trading with Syria

Phil returned to Baghdad at the end of July for a short time before heading to D.C. and then to a CENTCOM meeting in Tampa during early August. The new CENTCOM commander, General John Abizaid, had scheduled a conference for all key Iraq directors. Phil covered the oil sector. The long gas lines and shortages of products were prevalent in Baghdad all during August. Any insurgency group could see the effectiveness of attacks on oil pipelines in creating instability and insecurity.

The 30–50 trucks a day that Syria could have provided would have been a big help, but Phil declined to consider it. He said that he refused to embarrass the Bush administration. Even though I did not quite understand his comment, he made it clear that he did not want to discuss it. I suspect that it had something to do with the administration's efforts to get Congress to vote for economic sanctions against Syria.

In early August, Major General David Petraeus asked for my support to allow the 101st Airborne Division to exchange crude oil in northern Iraq with Syria for petroleum products. One of his soldiers had been killed that day while trying to control the crowds gathered to buy scarce products at a gas station. Petraeus told me that an Iraqi walked up behind the soldier, placed a pistol between his helmet and flak jacket, and blew his brains out.

General Petraeus needed gasoline, diesel, and cooking fuel for the Iraqi population in Mosul. My orders from Phil were very clear. I could not support Petraeus's request, but my stomach was in knots over the soldier's death. The only guidance that I could give Petraeus was to remind him of our West Point plebe year lessons on the responsibilities of a commander to protect his troops. That responsibility granted him more inherent authority than I could get him. As the commander, he could make a command decision in the field. Who could criticize an American commander for making a decision that would save the lives of his soldiers? Answer: the neocon civilians at the Pentagon who had never worn the uniform of the U.S. military.

Reports came in to the office a couple weeks later that the governor of Nineveh Governorate, the northern province containing Mosul, had signed a deal with the Syrians to provide crude oil to Syria in exchange for the delivery of 30–40 tanker trucks a day of products. The smile on my face was obvious. The person giving the report insisted that the CPA oil team take action to stop the sale. "I have no authority over governors," I said, and walked away. I had no energy or will to stop a governor's action that I knew would save the lives of U.S. soldiers. I wondered whether the young soldier who had been killed a couple weeks earlier would still be alive if only we had allowed SOMO to talk with Syria.

The exchange deal with Syria was quite clever. There was a small (8 inches in diameter) export pipeline near Sufaya, northwest of Mosul, for exporting crude oil to Syria. It was west of the large 46-inch export pipeline through Turkey. The 101st had already instructed the local oil ministry management to ship a small amount of crude oil across the border to Syria in exchange for electric power. This cross-border exchange for electricity had been going on prior to the war. Ambassador Bremer gave electricity advisor Steve Browning and me the approval to help the 101st negotiate exchanges with Turkey and Syria to get the electricity earlier in the summer.

There was unused capacity on the pipeline, and there was plenty of additional crude oil production capacity in the oil field at Ain Zalah. The local oil production management had the capability to ship much more crude oil to Syria than they were exchanging for electricity. The governor had signed a deal to send this additional crude oil in exchange for gasoline and diesel. It was a win for everyone. The Syrians got crude oil, the Iraqis in the Mosul area received products, and the 101st did not lose any more soldiers guarding the gas lines.

Not quite. There was an apparent loser: our civilian neocon leaders back at the Pentagon. The phone in the CPA oil office rang at 4 p.m. It was Mike Makovsky at the Pentagon.

Mike asked me whether I had given permission for the governor to sign a contract with the Syrians for oil.

"No, Mike," I responded.

Mike challenged me by asking who had given the governor permission.

"Mike, I don't know." I suspected that Petraeus had directed the governor, but I did not know and had no desire to investigate.

Mike had heard a rumor that Petraeus might have given the governor permission, and he asked me to confirm it.

"Mike, I told you that I do not know who authorized the governor."

Then Mike, irritated by my response, raised his voice. He said that if he confirmed that Petraeus directed the governor, Petraeus would be one sorry general.

I tried to calm him down. "Mike, please come out here to understand…" I heard a clang, and the phone went dead.

A week after Mike's threat about Petraeus, an article showed up in the *Early Bird Brief* — a popular roundup of global news reports on military, defense, and security matters — quoting a congressman who was demanding that the Pentagon fire Petraeus for trading oil with the Syrians. My immediate reaction was that Mike's threats, to yield that amount of power, must have been coming from someone higher up in his chain of command. Someone with good access to Capitol Hill was involved. Who, I did not know, but I have come to speculate that it was his AIPAC network.

The attack only showed up in the *Early Bird* one time. It had no legs, no follow-up. It may have been supplanted by a *New York Times* article on September 4, 2003, by Michael Gordon. In it, Gordon praised General Petraeus and the 101st for a job well done in establishing cross-border trading with Syria in several goods, not just oil.[3]

Instability and Death Count

Some congressmen started challenging the wisdom of their own Syria policies, but the Syria sanctions still passed with H.R. 1828.[4] Nick Rahall II of West Virginia stated on the floor that "such sanctions will not help alleviate the incessant attacks that our soldiers are facing daily in Iraq." Rahall also entered into the *Congressional Record* a *Wall Street Journal* article — "Iraq Adds Complexity for U.S., Syria as Washington Sanctions Damascus, American Troops Seek Syrian Trade Partners"[5] — that identified the disconnect between the reality of what was happening on the ground in Iraq and what Congress was pushing through as legislation.

The inquiries that I received from Makovsky during the summer of 2003 concerning the Haifa pipeline were bizarre. The neoconservatives' intolerant predisposition against trading with Syria for electricity and oil products was counterproductive at best. It was a distraction from our mission, and it cost American lives. It is naïve to believe that Makovsky was a standalone eccentric, as one Republican political appointee at the Pentagon wanted me to believe. Another political appointee referred to him as being part of "Wolfie's Gang," referring to Paul Wolfowitz. Makovsky had powerful sponsors who eventually rewarded him for his work. Since 2013, he has been the CEO of JINSA, with current compensation of over $600,000 per year.[6] According to Grant F. Smith's book *Big Israel: How Israel's Lobby Moves America* (2016), JINSA has accomplished for Mossad with the U.S. military what our CIA has tried to do for years by gathering intelligence on foreign militaries.[7]

The July 2003 pipeline attacks that shut down the refinery in Baghdad demonstrated two important realities. The first was that attacking pipelines was an easy way to create instability. This tactic would be used extensively by future insurgents: an estimated 3,000 attacks on pipelines took place during the five years between July 2003 and May 2008.[8] Many of those attacks were on the Iraq-Turkey (IT) export pipeline that exported oil to the Mediterranean. Some of that oil went to Israel after October 2003.

Second, the futility in opening a controversial pipeline to transport Iraqi oil to Israel became apparent to the folks who had been naïve to think that it was possible. The Iraqis demonstrated their willingness to use violence to stop this activity. The neocons needed an alternative plan, and they were going to execute it within a few months, by the end of 2003.

We lost 48 military personnel in July and another 35 in August, bringing the total since March to almost 300. We had no idea in the summer of 2003 that the KIA number would get to 4,489 by the end of 2011.[9] I have often wondered whether the number of KIAs was even important to people like Paul Wolfowitz and his group of neocons at the Pentagon. The answer became obvious on April 29, 2004, when Wolfowitz testified before the House Appropriations Subcommittee. In his answers to subcommittee members, he significantly understated the number of KIAs in Iraq. This number was updated every morning at the Pentagon and posted for anyone interested, so Wolfowitz could easily have checked on that number before testifying before Congress. He must not have been interested.[10]

Chapter 4:
The Glencore Plan – A New Strategy

*Iraqi Kurdistan replaces Iran
as primary supplier of oil to Israel.*

The long, hot summer of 2003 came to an end, and September brought a lot of personnel changes to the oil sector. Ahmed Chalabi* was appointed to a senior leadership position by Ambassador Bremer on September 1. Now president of the Governing Council of Iraq, Chalabi negotiated with Bremer to get his associate Ibrahim Bahr Al Ulloum appointed as the new oil minister. My boss Phil Carroll and I were scheduled to leave Iraq at the end of September. Phil's replacement arrived in late September, but my replacement declined his appointment, so I agreed to return to Iraq in October until a new replacement could be found.

The Coalition Provisional Authority (CPA) oil team had just responded to an inquiry initiated by Paul Wolfowitz at the end of August about oil being sold to Israel.[1] The inquiry came to me via an unclassified email (see Appendix A) from intelligence resources that asked how much oil was being sold to Israel. When I responded that no oil was being sold to Israel, the inquirer asked whether the Iraqis would know of any oil being delivered to an Israeli port. I responded that the only way to determine location of shipments was to track cargoes via *Lloyd's List* Intelligence.

I did not realize at the time that the wheels were now again in motion to get Iraqi oil to Israel by the end of 2003. As the Haifa pipeline plan was not possible, an alternative plan was being implemented. A CPA policy that prevented oil from being sold to oil traders was now being ignored in order to contract the company that had supplied as much as 90% of Israel's oil in the previous three decades. That company, originally Marc Rich & Co. AG, changed its name to Glencore in 1995.[2]

---

* Ahmed Chalabi, the Oil Ministry, and the CPA policy are also covered in Chapter 12 of *Iraq and the Politics of Oil*.

Chalabi's Oil Ministry

In September, I had my first meeting with the new oil minister in his offices in Baghdad. This meeting should have given me a clue that a new agenda was being executed. "Your Excellency, when I saw you in Washington, D.C., last January, did you ever think that we would be sitting here like this, nine months later?" I asked. Minister Bahr Al Ulloum said that he knew all along that he would get the oil minister job; it had been predetermined. I was curious: Why had he been so confident that he would become the minister of oil? What did he know that I did not know? I would eventually learn that Ahmed Chalabi backed Bahr Al Ulloum. He was Chalabi's man. This was part of a hidden agenda that I was not privy to.

The very talented Thamir Ghadhban remained in his position as CEO, but Bahr Al Ulloum made every effort to limit his responsibilities. We would learn over time that Bahr Al Ulloum saw Ghadhban as a professional and political threat, nothing more. Bahr Al Ulloum could have really distinguished himself in the oil minister role, but in the end, he failed to accomplish much of anything besides Chalabi's agenda.

The first key position that Bahr Al Ulloum replaced was that of the director general of the State Oil Marketing Organization (SOMO), then occupied by Mohammed Al Jibouri. Al Jibouri was summoned to meet Chalabi at a hunting club in Baghdad that also served as the headquarters of the Iraqi National Congress shortly after Bahr Al Ulloum became the oil minister. Al Jibouri was instructed about how he would operate the SOMO organization going forward, especially in the issuance of contracts for selling crude oil.

Al Jibouri was told to sell oil to Glencore, the international commodities company. Founded by Marc Rich in 1974, this trading company had supplied most of Israel's oil during the previous three decades.[3] Al Jibouri objected. He would not allow sales to trading companies during his watch. It was a CPA policy not to sell to the oil trading companies and to sell only to end users, those who owned refineries. Al Jibouri fully supported that policy because he had seen the corruption that oil traders fomented during Saddam's regime. A 2004 Charles Duelfer investigation, for instance, reported that Glencore paid more than $3.2 million in "illegal surcharges" (i.e., bribes) to the Saddam government over the course of the UN's oil-for-food program before 2003.[4] Chalabi warned Al Jibouri that if he did not do as he was told, he would be replaced.

The Glencore Plan – A New Strategy

Ibrahim Bahr Al Ulloum removed Al Jibouri as the SOMO lead and replaced him with someone who would follow instructions from Chalabi's team. SOMO started selling to Glencore and other traders. Why was it so important to Chalabi that Glencore become a customer of SOMO? It was Chalabi delivering on his prewar promise to the Israel lobby: to deliver Iraqi oil to Israel.

I was in the U.S. in early October 2003 when Al Jibouri was removed from the SOMO position. Upon returning to Iraq in mid-October, I immediately requested a meeting with Ambassador Bremer to discuss the removal of Al Jibouri. My meeting with Ambassador Bremer was disappointing.

I entered Bremer's office, and he welcomed me back to Iraq from my two weeks away. I mentioned to him that in June he, Phil Carroll, and I had made it a policy that oil sales would be made only to companies with refineries — end users — and not to third parties such as oil traders or brokers. I mentioned that Mohammed Al Jibouri had been replaced because he was enforcing that important CPA policy, but Ambassador Bremer's response took me by surprise. He simply said that we could not eliminate corruption from Iraq during our watch.

I walked out of Ambassador Bremer's office wondering what had just happened. Was corruption now being allowed? Did I misunderstand the ambassador? It took me several years to conclude that the ambassador was following orders, nothing more. Bremer reported directly to Secretary Rumsfeld in 2003, but Rumsfeld knew little about oil. His deputy was the oil expert on his staff — those orders had come from Deputy Secretary Paul Wolfowitz.

Glencore and other trading companies started receiving Iraqi oil in October 2003 (and continue today). The oil was primarily delivered to Glencore by SOMO at the port of Ceyhan, Turkey, at the western end of the Iraq-Turkey (IT) export pipeline. From there, Glencore could deliver the oil to the Israeli ports of Haifa or Ashkelon. The IT pipeline was attacked many times during the second half of 2003 and the following years to stop the export of about 500,000 barrels a day to the Mediterranean. We did not fully comprehend back then why that pipeline was so constantly attacked, but I suspect today that the Iraqi attackers knew that SOMO was selling their oil to Israel and that the Iraq-Turkey pipeline was the primary export route. I remember what the Iraqi director general told his men in July 2003: "The

Americans were not sending our oil to Israel. If they were, I would help you blow up our pipelines."

The Kurds

Sales directly from SOMO to traders such as Glencore and Trafigura continued sporadically from 2003 until an arrangement more favorable to Israel was executed in 2013, when Israel leveraged its close relationship with Iraqi Kurdistan. In 2014, the Kurds constructed a new 36-inch pipeline that started delivering Kurdish oil through the Iraq-Turkey pipeline to the port of Ceyhan. Their pipeline route bypassed the many areas of northern Iraq controlled by those Sunni tribes who had been attacking the pipeline since late 2003. The Kurds created a stir with these shipments of oil to the port of Ceyhan. Many believe that it marked the beginning of an independence movement by the Kurds from the rest of Iraq.[5] This oil export activity took place without the approval of the Iraqi federal government, so Baghdad considered all Kurdish exports as illegally smuggled Iraqi crude oil.

The Kurds could not sell their oil to any country officially recognized by the Iraqi government, including the United States. The Kurds had in fact tried to ship cargoes to the U.S. Gulf Coast, only to be threatened with a seizure order, through the U.S. courts, from the Iraqi federal government. One ship (*United Kalavryta*) was able to evade U.S. Customs in 2014 and eventually delivered its cargo to Israel.

Israel was the only Middle East country able to help the Kurdistan Regional Government (KRG) with this international smuggling operation because they had no diplomatic relations with Iraq. Since Israel's founding in 1948, Iraq has never recognized Israel. That policy seems likely to continue. In May 2022, the Iraqi parliament outlawed normalization through a law, approved by 275 out of 329 seats, titled "Criminalizing Normalization and Establishment of Relations with the Zionist Entity."[6]

The volumes delivered to Israeli ports were much more than Israel needed, forcing them to export the excess oil to other countries, without the original country of production documented. Unfortunately, there are no trade statistics of any oil imports shared by the Israeli government. Both Glencore and Trafigura oil trading companies played a role in this operation. They both trace their roots to Marc Rich & Co. AG, and all three oil trading companies have had close links to the Israeli government.

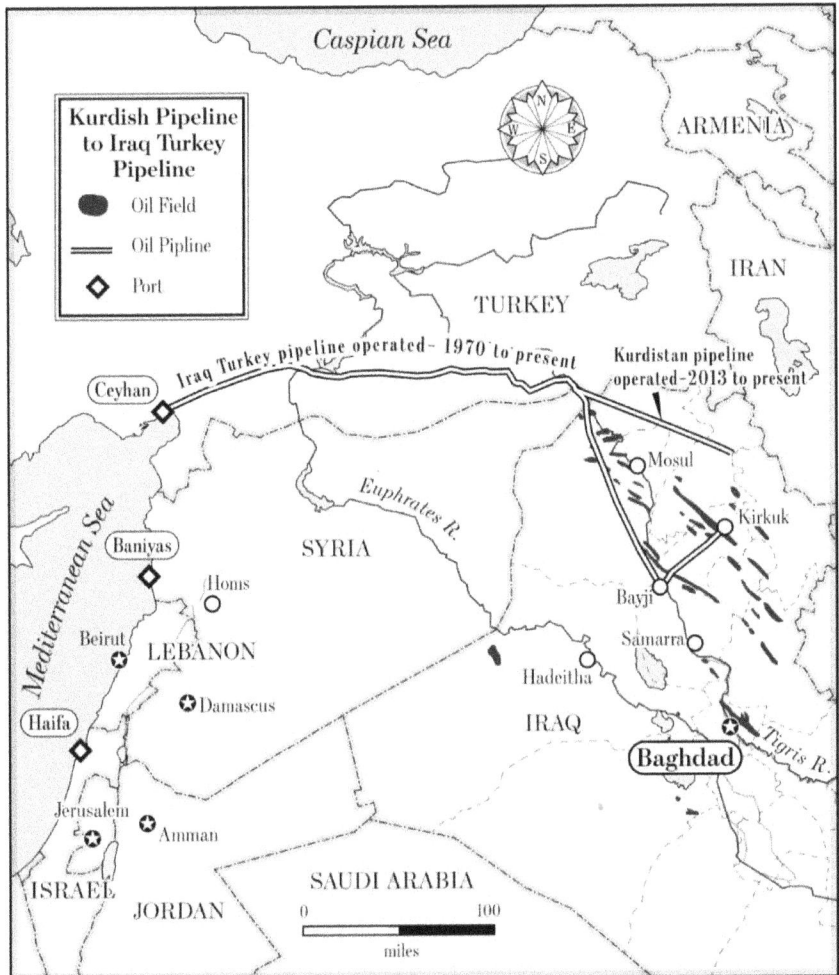

Figure C: Pipeline showing the new 2013 oil route from Kurdistan to Ceyhan. (MapArtist.com)

Since 2014, many attempts have been made to negotiate an oil agreement between the KRG and Federal Iraq under the 2005 Iraqi constitution, but most have failed. Baghdad and the Kurds disagree on the interpretation of their constitution. In the meantime, Israeli energy security has flourished, based primarily on this Kurdish oil production of 450,000 barrels a day. Israel has received a constant supply of oil at a low price.

In February 2022, the Iraqi Federal Supreme Court ruled against the legal foundations of Iraqi Kurdistan's independent oil sector. As Ben Van Heuvelen and Ben Lando of the *Iraq Oil Report* explain:

The court's decision could force a seismic shift in the balance of oil powers between the federal government and the semi-autonomous KRG — potentially upending a 450,000 barrel per day (bpd) market and altering the dynamics of a government-formation process that has seen multiple blocs court Kurdish MPs.[7]

This ruling immediately had an impact on international oil companies doing business with the Kurds. For instance, according to an August 2022 *Iraq Oil Report* article, the only American company facing potential lawsuits so far was a small, privately held Dallas-based company, HKN Energy Ltd. The Iraqi federal government issued a letter invalidating their contract with the Kurds. As of August 2022, large international oil companies doing work in KRG — including Chevron, TotalEnergies, Gazprom Neft, and Rosneft — have not yet been sued.[8] The two Swiss-registered trading companies Glencore and Trafigura should also be obstructed by this decision. Will this decrease Israel's access to Iraqi oil? I suspect it will.

Israelis in the U.S. Government

Any negative impact on Israel's energy security has previously prompted action from the U.S. government. What will the U.S. response be this time? From August 2021 to May 2023, the Special Presidential Coordinator for Global Infrastructure and Energy Security was one Amos Hochstein.[9] Mr. Hochstein was born and raised in Israel, and he served in the Israeli Defense Forces (IDF), as most Israeli citizens perform military service. He moved to the U.S. and received appointments to government positions during the Obama and Biden administrations.[10] Any official policy from the Biden administration will most likely require Mr. Hochstein's involvement.

Why has the Biden administration appointed an Israeli citizen and veteran of the IDF to a critical U.S. government energy policy position? Excellent question. I asked the same question in 2005 when I learned that the Bush administration had appointed an IDF veteran and suspected Mossad asset to an important energy policy position at the Pentagon in 2002. I now know that that person was very seriously conflicted. I know that U.S. soldiers were killed in Iraq because of policies pushed by that seriously conflicted person.

I refer, of course, to Michael Makovsky. The following came from an interview* in the *St. Louis Post-Dispatch* (see Appendix B) from February 20, 1989, just before he emigrated from the U.S. to Israel:

> After his graduation from the University of Chicago, he worked for then-Prime Minister Shimon Peres. "I would write weekly news summaries on what United States and British newspapers were saying about Israel in general and Peres in particular," Makovsky noted... Makovsky also wrote media analyses about the Pollard spy case, which strained relations between the United States and Israel two years ago. Pollard currently is serving a prison term in the federal penitentiary in Marion, Ill.†
>
> Makovsky's immediate superior during this period was Uri Savir, now consul general for Israel in New York. Savir served as an intermediary between Makovsky and Peres and, last November, encouraged Makovsky to enter the Israeli foreign service....
>
> "But I have strong feelings about helping to build a Jewish state. There's an excitement about living in a young country that's only 41 years old, a country with problems on its borders. It's like returning to your roots." Under Israel's Law of Return, Makovsky immediately will enjoy dual citizenship....
>
> He doesn't plan to renounce his U.S. citizenship. His experience and background with both the Israeli and U.S. governments, Makovsky said, should operate in his favor. The foreign service exam is a seven-month process, and he concedes that his move to Israel "is not practical and was not an easy decision."[11]

I refer to Makovsky as a suspected Mossad asset because an intelligence resource advised me that 98% of the Israeli foreign service are Mossad. Unless I were to see verified cables under Makovsky's signature as an Israeli diplomat, then he was Mossad. Confirmation of his Mossad status was impossible, but Makovsky provided me an opportunity to ask him about it in December 2005. We were both at a meeting at the State Department when the attendees were all notified that the guest speaker was delayed for 30 minutes. Makovsky approached me about a personal issue, and when he finished, I took the opportunity to question him about the facts presented in his 1989 *Post-Dispatch* interview concerning his service in the Israeli Defense Forces and Israeli foreign service. His response to me was that he would

---

* The article is rather difficult to find on the Internet through normal search engines. The article in its entirety, as well as its archive link, has been reproduced in Appendix B.
† Pollard was paroled in 2015.

neither confirm nor deny such accusations. I gave him the opportunity to deny the facts in the *Post-Dispatch* article, and he refused.

Makovsky's appointment to the DOD policy job in 2002 was a conflict of interest under the guidelines of the Foreign Influence paragraph (Guideline B) of the National Security Adjudicative Guidelines.[12] His work with a foreign entity was an obvious conflict of interest that should have been disclosed to every American working with him. If I had known then what I know today, I could have taken action to mitigate his activity and saved lives; or, I would have resigned.

Chapter 5:
Israel's Energy Security – The Motive

*Marc Rich extended the Iran–Israel
oil supply from 1979 until 1995.*

In 2003, Joseph Paritzky, the Israeli Minister of National Infrastructures at the time, confirmed that, beginning in the late 1990s and continuing until October 2003 — when they began receiving Iraqi oil via the pipeline to Ceyhan, Turkey — Israel had been hurting from having to pay a huge premium on oil imports. A brief history of Israel's energy security is in order.

The International Energy Agency (IEA) defines *energy security* as "the uninterrupted availability of energy sources at an affordable price."[1] The last Shah of Iran Mohammad Reza Pahlavi and oil trader Marc Rich played important roles in the energy security history of Israel. Israel imported 60–90% of its oil from Iran before 1979.[2] The Iranian Revolution in 1979 overthrew Pahlavi, placing Israel's energy security at extreme risk. The new regime of Ayatollah Khomeini changed all the Iranian oil sales contracts to prohibit further oil sales to Israel. The supply of oil to Israel from Iran was even temporarily halted in 1979. Marc Rich played an important role in rectifying that by winning a continued flow of Iranian oil to Israel until 1995.[3]

Top-Secret Pipeline

Daniel Amman authored a biography of Marc Rich, *The King of Oil* (2009), a few years before Rich's death. This book revealed many well-kept secrets that surprised many in the international oil industry. Javier Blas and Jack Farchy authored a recent book, *The World for Sale* (2022), that reinforces the facts in Amman's book and further sheds light on the secretive world of Marc Rich and other oil traders. Even those outside of the oil industry knew that Rich was a very successful oil trader for over 20 years, but how he succeeded was a heavily guarded secret before these two books were published.*

---

* Marc Rich and his role in delivering oil to Israel are discussed in Chapter 12 of *Iraq and the Politics of Oil*.

Amman shared with me the two things that most surprised him when he interviewed Rich and others during his research. The first was Rich's close relationship with the head of Mossad. Rich kept the head of Mossad's phone number on his speed dial, and the two gentlemen spoke several times a week. The second revelation was the so-called "top-secret pipeline" that ran from the Israeli port of Eilat in the south to the port of Ashkelon on Israel's Mediterranean coast.

The secrecy around the "top-secret pipeline" was quite real. I worked for Mobil Oil in Saudi Arabia for over four years during the late 1980s, and it came as a surprise to me when I first learned of its existence after reading Amman's book in 2015. I contacted a former Mobil Oil executive who traded Middle East oil for over two decades. Even he was not aware of the pipeline's existence.

The pipeline, owned by Iran and Israel via the joint venture Eilat-Ashkelon Pipeline Company (EAPC), was constructed in the late 1960s to transport Iranian oil to both Europe and Israel. It is a comparatively large pipeline, with a diameter of 42 inches, capable of transporting a million barrels a day. Israel took sole control of EAPC in 1979 after the Iranian Revolution. A Swiss court ruled in 2016 that Israel owed Iran $1.2 billion for its share of the pipeline, a sum that Israel refuses to pay.[4]

The Israeli government continues to value the secrecy of the EAPC pipeline for their energy security. In late 2017, the Israeli Knesset passed a law ordering up to 15 years in prison for anyone leaking confidential information about the pipeline. The Israeli law states that releasing operational information about the activities of the pipeline company is an act of espionage.

How Marc Rich Got Rich and Then Became a Fugitive

This secret Israeli pipeline was a major source of Marc Rich's fortune from 1973 through the end of 1994.[5] Rich's role in getting the oil flowing again from Iran to Israel in 1979 was recognized and appreciated throughout the Israeli government. The following quote from Daniel Amman's biography of Rich highlights this fact:

> Israel's salvation came from none other than Marc Rich, a fact that has remained largely unknown to this day. "Israel owes a great debt to Marc. He provided Israel with all its energy needs in its most difficult time," Abner Azulay told me. Azulay, a former Colonel in the Israeli defense forces and high-ranking Mossad agent with a solid

network of political contacts, today directs Rich's philanthropic foundation.[6]

Rich had also developed excellent business contacts among the professional technocrats in Iran. The professionals just wanted to sell their oil and get paid for it — strictly a business transaction. This business relationship temporarily received priority over politics until the mid-1990s, when the supply arrangement between Iran and Israel ended and Marc Rich sold his company.[7]

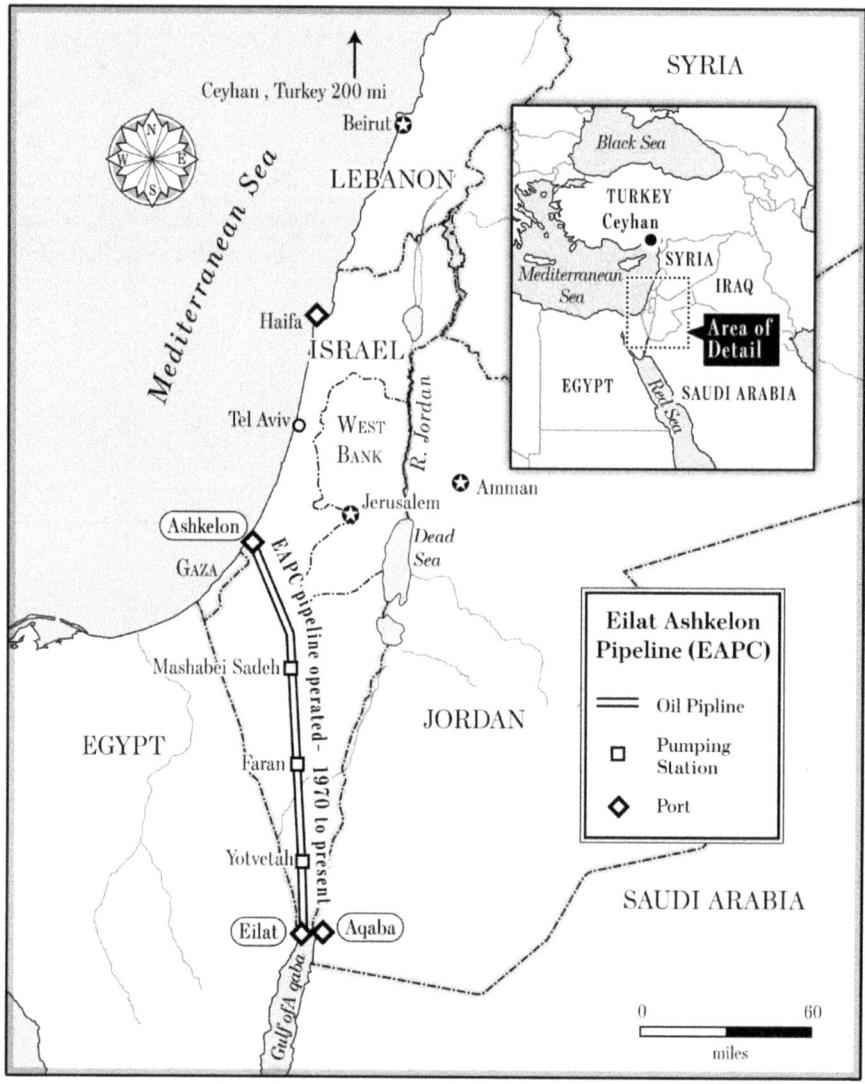

Figure D: Top-secret EAPC pipeline in Israel. (MapArtist.com)

Marc Rich & Co. AG experienced booming business during the 1970s and 1980s. The Swiss-registered company expanded into several new countries and into various commodities. Rich sought out opportunities in whatever countries were sanctioned by the U.S. but not by Switzerland. Iran was not the only sanctioned country where Rich did business. Others included Cuba, Angola, apartheid South Africa, and Nicaragua under Daniel Ortega.[8]

The 1990s proved to be more of a bust than boom decade for Rich. A few poor deals in his non-oil commodities business cost his company several million dollars. The biggest single bad deal took place in 1992. Marc Rich & Co. AG tried to corner the zinc market in a very speculative trade that cost him $172 million.[9]

The strain of his 1980s criminal conviction in the U.S. for income tax evasion was having a negative impact on his ability to conduct business. He was labeled at the time as an international fugitive by the U.S. government, particularly by the FBI, and was so labeled throughout the world. Having been convicted in the U.S. District Court of the Southern District of New York in a high-profile case prosecuted by Rudy Giuliani, Rich was wanted on charges including income tax evasion, racketeering, and trading with the enemy. He faced a maximum penalty of 325 years in prison and/or $100 million in fines.[10]

President Bill Clinton pardoned Marc Rich on his last day in office (January 20, 2001) in what William Safire of the *New York Times* called "the most flagrant abuse of the presidential pardon power in U.S. history."[11] Several former Israeli prime ministers had lobbied the Clinton administration to get Rich pardoned. Clinton credited the testimony of I. Lewis "Scooter" Libby, Rich's longtime lawyer, as having a huge influence on his decision.

What role did Marc Rich play after his pardon and in the months preceding the Iraq War? Was he advising his former lawyer Lewis Libby during 2003, while Libby provided guidance to our planning team at the Pentagon and communicated with Ahmed Chalabi in Baghdad? Did Marc Rich owe a debt to Libby for helping with his pardon from Clinton? I have often wondered about the answers to those questions. I asked Daniel Amman these same questions, but he replied that his information about Rich ended on the date of his presidential pardon. Marc Rich died in 2013, so those questions will probably never be answered.

Israel Is Doing Great Now

Once the supply arrangement with Iran was terminated in 1995, the sources of oil to Israel varied. Most of the volumes came from Russia and any other sources where Israel's oil traders could get it. The long-term supply arrangement with Iran via the EAPC pipeline had secretly provided benefits to both Iran and Israel for over 25 years. It was an important non-Arab route for moving Iran's oil to customers in the Mediterranean, bypassing the Suez

Canal as well as the Sumed pipeline in Egypt. For Israel, it provided a constant supply of oil through a high-capacity pipeline that added efficiencies to its energy system. This loss was significant.[12]

The 25% energy cost premium that began in 1995 provided a huge motive for seeking other inexpensive solutions to Israel's energy security issues. Iraqi oil has provided a catalyst for a booming Israeli economy since 2003. Israel's economy today is one of the strongest on the planet, based on several measures.

For instance, according to the World Bank, Israel's per capita GDP in 2021 was $51,430. That is higher than that of Germany, the United Kingdom, France, Japan, the United Arab Emirates, Kuwait, and Saudi Arabia. Another measure is the number of billionaires as a proportion of the population. Israel has done much better than the U.S. in creating billionaires since the Iraq War of 2003. According to *Haaretz*, Israel had 8 billionaires in 2003, growing to 128 billionaires in 2019, among a population of 8.5 million Israelis.[13] The number of U.S. billionaires in 2020, according to *Forbes* magazine, was 630, among a population of 340 million.[14] Israel has 15.06 billionaires per million people, while the U.S. only has 1.85 billionaires per million people. Israel's per capita number of billionaires is therefore more than eight times that of the U.S. Have lower oil prices since 2003 contributed to the tremendous economic growth of Israel? They have certainly helped.

U.S. oil imports from Iraq are negligible. The Energy Information Agency (EIA) reported in 2022 that Iraq supplies the U.S. with about 1% of its oil demand. Iraq supplied the U.S. with more oil during the sanction years just before 2003 than in any year since 2003.[15] Most of Iraq's Persian Gulf oil exports currently go to East Asian countries. According to the Observatory of Economic Complexity (OEC), 38% of Iraq's exports* in 2020 went to China, 29% to India, and 7% to South Korea. None of those countries participated in the 2003 war.

I shared some of this data with a retired federal judge from the Midwest a few years ago while I was giving a presentation in Florida. The judge seemed quite interested in my explanation, but then admitted to missing the point I was making. "What harm was there in helping our friend Israel with getting cheaper oil for their economy during our invasion of Iraq?" the judge asked. The judge had missed two critical points: first, that the

---

* The OEC statistics do not include the Ceyhan exports to Israel by the Kurds.

neoconservatives were the primary drivers for war with Iraq, and second, that cheap Iraqi oil for Israel was their primary motivation. I responded to the judge's question by stating that I had nothing against Israelis getting cheaper gas at the pump; I just did not think that 4,489 U.S. soldiers and over 250,000 Iraqis should have had to die to enable it.

Chapter 6:
The Suspected Mossad Asset

*JINSA/Mossad succeed in
making the U.S. military use an old CIA blueprint.*

Why would the Bush administration secretly place a suspected Mossad asset in a senior Middle East oil policy position in the Office of the Secretary of Defense (OSD)? Why would the Bush administration assign a top-secret sensitive compartmented information clearance (TS SCI) to a suspected Mossad asset at the objection of the DOD security clearance professionals? Why was this suspected Mossad asset given the authority to make decisions and direct activity that resulted in casualties to U.S. soldiers? These were all questions I asked in the summer of 2005 when I discovered Michael Makovsky's background. Any American, if he had experienced what I did in 2003, would have asked the same questions about what I believed to be a gross conflict of interest. Let me take you back to the summer of 2003 and my relationship with a senior statesman of the U.S. oil industry.

Phil Carroll, Retired CEO of Shell U.S.

It was the second week of May when Mr. Philip J. Carroll arrived in Baghdad to take the senior U.S. government oil position under the CPA, reporting directly to Ambassador Bremer. Phil was my new boss. A retired CEO of Shell Oil and retired chairman of Fluor Engineering, he had many excellent contacts in the Houston oil patch. He was a friend of former President George H. W. Bush and his wife Barbara. Phil Carroll had political insight into the current administration that I lacked. I always depended on and deferred to his political judgement, especially with some of the new decisions coming out of the CPA leadership under Ambassador Bremer.

Phil asked to meet with me the first morning after his arrival for a brief private meeting of introduction. I had met Phil at the Pentagon a few months earlier and admired his direct approach. I was nevertheless surprised at the first words he said to me: "Gary, if I see anything close to an oil grab out here, then I will immediately resign and catch the next plane back to the U.S. Do you understand me?" The Western press was convinced that this war was being fought so that Bush/Cheney could provide Iraqi oil to their friends in the U.S. oil industry — an oil grab. I welcomed his comments and replied that I had not seen anything close to an oil grab, but that if I did, then I would also submit my resignation and we would depart from the country together.

Phil and I shared a common background, besides several years of experience in the oil industry. We were both commissioned as Army officers after college. Phil spent several months in the Army after graduating from Tulane, and I spent several years in the Army after graduating from West Point. We both shared an appreciation for the soldiers with whom we were serving, and we felt insulted by the focus of the press, that this war was all about oil. Seeing U.S. soldiers killed to benefit one or a group of oil companies was something that both of us found repugnant.

From April to September 2003, the Iraq oil sector went from the best of times to the worst of times. The first oil production started on April 23, several months ahead of schedule. We had a formal plan of oil projects agreed upon in June by the U.S. Army Corps of Engineers, the Iraqi oil ministry, and the CPA. Oil export sales started in June from the Iraqi ports in Basra to the south and from the port in Ceyhan, Turkey, to the north. Competent Iraqi oil leadership was appointed to run the oil ministry. It appeared that our work in the oil sector could be easily managed going forward.

Phil Carroll and I made a CPA staffing plan that included the winding down of our CPA oil mission by the end of 2003 or early 2004. Phil scheduled the repatriation of experienced Department of Energy personnel during June and July and scheduled his extended absence from Iraq during July with a return to Baghdad in August. His final departure and repatriation were planned for September. My role was to be his deputy. In hindsight, the months of April, May, and June proved to be the calm before the storm, and we were not staffed to weather that storm.

Clues of Incompetence and Corruption

Several things occurred in July that seemed unrelated at the time but appear very much related today. The first was Michael Makovsky's insistence that the Iraqi CEO and I figure out the status of an old Iraqi pipeline to Israel. He said that the status of this pipeline was the highest priority of the senior civilians at the Pentagon (probably an outright fabrication, with the possible exception of Deputy Secretary Wolfowitz, who always showed an interest in oil issues). This was followed by attacks on important pipelines that supplied crude oil to the refinery in Baghdad. These attacks immediately caused the severe shortages for Iraqis, as I have previously described.

Next was the refusal of Makovsky and Pentagon civilians to support sourcing supplies of gasoline and diesel from Syria. This policy further exacerbated the shortages and encouraged attacks on U.S. soldiers. Frank Miller, a senior member on the NSC in 2003, stated in a videoconference that the press reports hitting the U.S. newspapers made it appear that we were losing control of Iraq. The front-page newspaper photos of gas lines several miles long in 110-degree heat were indeed accurate.

The CPA budget director, Retired Admiral David Oliver, only further compounded the problems by his refusal to pay the Iraq oil sector workers adequately.* Oliver's miscalculations resulted in massive labor strikes throughout the oil sector during July and August. Oliver, who had arrived in early May 2003, only stayed until late 2003, when he returned to the U.S. for a senior position with a large defense contractor.

Oliver was close to Senator Joe Lieberman, and that was supposedly how he got the job in the CPA. Oliver loved to brag and share stories about his close relationship with Admiral Hyman Rickover, the "Father of the Nuclear Navy." During our 2005 "lessons learned" project for the Army, Phil Carroll and I identified Oliver as the biggest impediment to the completion of our oil sector mission in 2003, thanks to his failure to pay the oil sector employees adequately and also to his failure to fund the oil projects agreed to in June by the CPA.

Why did Oliver refuse to pay the Iraqis adequately? Why did he fail to fund the CPA's oil projects? Was he part of the Israeli oil agenda or some other agenda? Oliver said that he needed to build a new civil service payroll system to replace the "pay-for-performance" system that Iraq had been using in the oil sector, even though a U.S. Army colonel assigned to the oil sector had studied this system and told me that it was an excellent payroll system. Oliver refused to listen to the colonel and insisted on bringing consultants from the States to design a new system. His consultants showed up in September. One of the consultants was a Mrs. David Oliver. I immediately went to Ambassador Bremer's office to file a complaint for conflict of

---

* For more on Oliver's interference, see Chapter 9 of *Iraq and the Politics of Oil*.

interest on Oliver's part. Bremer's admin director told me that a waiver had been signed for Oliver's wife by the ambassador. Oliver's motivation appeared to be for personal reasons.

Phil Carroll returned to Baghdad for a week in late July before departing for short meetings in D.C. and the military headquarters in Tampa. We entered crisis management mode. It seemed that the priorities that Phil had shared with me on his first day in May had changed. When I challenged him on the bizarre Syria decisions by the Pentagon, I sensed that something had changed in Phil's attitude. In hindsight, I suspect that he had learned of the Israel oil agenda, but he refused to embarrass the administration by resigning and departing from Iraq.

Phil did later confide in me, in September 2003, that he regretted coming out of retirement to take the Iraq job. He told me that he should never have answered the phone call from Secretary Rumsfeld earlier in the year to accept the job. I confessed to him that I was asked in February to recommend the best way to convince Phil to take the leadership position on the Iraq oil team. My response was: "I do not know how any patriotic American could refuse a personal request directly from the Secretary of Defense when their country was going to war." Phil then looked me in the eye and said, "So, you are the son of a bitch I should blame." We both got a laugh at his comment, but he was serious about his regrets. He departed from Baghdad in late September just after his replacement arrived. He could not wait to leave.

Phil Carroll out-briefed Paul Wolfowitz before returning home to Houston. Phil eventually delivered a few oil industry speeches praising the great service of our soldiers in Iraq while also criticizing the Iraq oil sector privatization plans of the senior civilians at the Pentagon. I continually reached out to Phil after his departure from Iraq to seek his guidance during the remainder of my initial tour and after returning to the U.S. in July 2004. For example, in 2005, when the Army asked me to write a "lessons learned" paper about our 2003 mission in the Iraq oil sector — just in case they had to do a similar mission in an adjacent Middle East country, like Iran — Phil was the first person who helped with that project.

I will never forget the guidance Phil provided me during the summer of 2005, when I discovered that Michael Makovsky was probably some type of an asset for Mossad. That telephone conversation with Phil Carroll was a real eye-opener for me to the realities of the tremendous power of AIPAC over the Bush administration and the country.

Back Home

My first tour in Iraq ended in July 2004 with my resignation as a DOD Senior Executive Service (SES) employee. I was the longest-serving American in Iraq at the time. Physically and mentally exhausted, I needed several months to recover. My physical location was at home in Virginia, but mentally, I remained in Iraq. My first activity every morning was to check all the news sources on the Iraq oil sector. I maintained email and telephonic contact in Iraq and with colleagues at several agencies in the D.C. area.

In the summer of 2005, one of those colleagues working at an intelligence resource sent me a Daily Kos article titled "'Outing' the Neocons: The Office of Special Plans."[1] My colleague mentioned that I might know the five names mentioned in the article: David Schenker, David Wurmser, Michael Maloof, Michael Rubin, and Michael Makovsky. I was surprised at the descriptions of the first four names, having met some of them at the Pentagon in 2002. However, I was horrified at what the article stated about Michael Makovsky.

Makovsky had experience working for senior Israeli government officials after college, followed by a job with Senator John Danforth of Missouri. As I detailed in Chapter 4, in 1989, Makovsky emigrated from the U.S. to Israel to join the Israeli Defense Forces (IDF) and later had a career in the Israeli foreign service. The *St. Louis Post-Dispatch* article which I cited in that chapter (see also Appendix B) highlighted how Makovsky left the U.S. to join the foreign service of a foreign country: "Makovsky feels strongly about moving to Israel, where he'll have to serve a stretch in the army before entering the country's foreign service."[2]

My thoughts went back to a specific day in November 2002. Makovsky entered our prewar planning Special Compartmented Information Facility (SCIF) with an obvious look of anger on his face. I asked him what was bothering him, and he replied that he had just come from his third interview with the Pentagon security group about his top-secret clearance. I had never met anyone who needed more than one interview to get a clearance, so I commented that he must have an interesting background. He responded that the security professionals did not like his Israeli friends. He planned to go to Under Secretary of Defense for Policy Doug Feith to fix his problem.

My thoughts then shifted to all the problems Makovsky had directly or indirectly caused during 2003 from his obvious conflict of interest. I sent the Daily Kos article to Phil Carroll's secretary in Houston and scheduled a phone call with him a day later. "Phil, were you familiar with Mike's background?" I asked. Phil admitted that he had learned of Makovsky's history during the summer of 2003. He informed me that Doug Feith had overruled the Pentagon security personnel to award Mike a top-secret clearance.

"Why did you not share that information about Mike with me in 2003 when you learned of it?" I asked Phil.

Phil asked me, "What difference would it have made? There was nothing to be gained by letting everyone know back then about Mike's background."

"Phil, why would our government secretly place a Mossad asset in a senior Middle East oil policy position in the Office of the Secretary of Defense? Why would our government assign a top-secret sensitive compartmented information clearance to a Mossad asset at the objection of the DOD security clearance professionals? Why was this Mossad asset given the authority to make decisions and direct activity that resulted in casualties to U.S. soldiers?"

My frustration and anger were clear in the tone in my voice. I had been deceived by my government. I felt betrayed by senior leaders at the Pentagon who had placed this highly conflicted individual with a group of patriotic Americans while keeping his background a secret. I had been duped into giving credibility to a person who should have been treated like any other foreign ally to our country. Any comments or requests from him should have been in writing and treated as though they had come from a foreign country.

Phil stated that I needed to be careful with the questions I was asking, and especially with my tone. He warned me that AIPAC would protect their people and come after anyone perceived as attacking them.

"Phil, I am just asking questions," I said. "But what will they do, kill me?"

Phil said that was not their modus operandi. He proceeded to share a detailed story of how I would pick up the *Washington Post* in my front yard and notice my name in bold print on the front page. The spelling of my name would be accurate, but everything else in the *Post* article would be fabricated. My character would have been successfully destroyed. He told me that any future retraction would appear on page 18 of the *Post* in small print. He told me that my good name would be destroyed.

Phil's description seemed so real, as though it had happened to him. "Phil, you sound like you had a personal experience with them," I said.

He told me that he had seen that exact scenario happen to others. When he was the CEO of Shell back in the 1990s, there was a downturn in the price of oil, where it reached levels under $12 per barrel. I was familiar with that period because at Mobil Oil we cut staff by more than 25%, as did most of the large oil companies. Phil and his vice president of human resources at the time decided that they needed to make major cuts in their funding of the Shell Foundation for U.S. charities. He signed a directive that decreased the funding for all charities by the same percentage.

An assistant HR director met with Phil before the letter was released for implementation to explain to him that he could decrease contributions to all charities except those that gave to Israeli causes. He warned Phil that there would be personal attacks on him for reducing the contributions to those charities. Phil amended his original directive to protect the contributions to the Israeli-affiliated charities.

"Phil, I am not a lawyer, but I think you just described a classic case of extortion," I said.

Phil concluded the call by explaining to me that he was giving me fair warning, a warning to a person he considered a friend. I would have to decide for myself whether to heed his warning.

We stayed in contact over the following years, and he provided great guidance to me whenever I asked. In 2008, he put me in contact with Steve Coll, the Pulitzer Prize-winning author, who was writing a book at the time about my old company, ExxonMobil. Coll asked whether I had any prewar contacts in preparation for the Iraq War as an employee at ExxonMobil. I assured Steve that my only prewar Pentagon contacts had come via my Army Reserve unit, and that ExxonMobil was not aware of my work until they read about it in the *Wall Street Journal* in April 2003. Coll's book, *Private Empire: ExxonMobil and American Power*, was published in 2013.[3]

Coll seemed most interested and amused by my story about the old Haifa pipeline to Israel. He asked me what the most bizarre thing was that happened to me during the summer of 2003. At first, I mentioned that a lot had taken place that summer, but nothing I considered bizarre. He asked again. I thought for a minute and then remembered how bizarre I thought it was that Michael Makovsky made a high-priority inquiry about a pipeline I had never heard of. I told Steve the entire story. Chapter 11 in Steve's book is titled "The Haifa Pipeline." Coll made it a point to interview Makovsky as well about the Haifa pipeline story. Makovsky commented that he was merely responding to an inquiry from the Israelis to get the status of the pipeline. But why was Makovsky responding to a foreign country's request? He was an analyst working for Doug Feith that summer.

Steve Coll was very helpful with a high-priority oil export project I was working in 2008. The U.S. ambassador and senior military commander could not get Hussain al-Shahristani, Iraq's minister of oil, to approve the funding for a strategic export project in southern Iraq. Time was running out, and the oil minister needed political help within his government to get their approval. Coll suggested and facilitated help from the international press to bring appropriate focus to the benefits that the project would bring to Iraq. Four years later, at the commissioning ceremony for the project, the oil minister declared the project to be the most important infrastructure project that Iraq had completed in decades. The $2 billion approved for the project in 2008 enabled record oil exports from southern Iraq and hundreds of billions of dollars in additional revenue during the last ten years (2013–23).

I eventually lost contact with Phil Carroll in 2013 and later learned that he had become seriously ill with cancer. He died in 2014.

JINSA

JINSA, the Jewish Institute for National Security of America, was described to me as closely linked to Mossad long before Makovsky became the CEO in 2013. According to JINSA's financial records for 2019, $432,000 was spent on a program to send about ten retired generals and admirals to Israel for about a week. The following from JINSA's website describes a program that has been fully funded by JINSA since 1981:

> More than 450 retired U.S. generals and admirals have participated [in] the Generals & Admirals Program, including service chiefs and combatant commanders. The cornerstone of the program is a trip to Israel, during which newly retired senior U.S. flag officers meet with top Israeli military, intelligence, and government leaders, including the Israeli Prime Minister, Israeli Minister of Defense, Israel Defense Forces (IDF) Chief of Staff, and Director of Mossad.[4]

JINSA is physically located in a building not far from the White House. Shortly after my book *Iraq and the Politics of Oil* was released, I was threatened by an individual whose LLC was registered under an address in the same building. A close friend and fellow West Point graduate recommended that I use a private investigator to assess the threat. My private investigator reported back to me that every company listed on the registry at the building was suspicious. The FBI was notified.

Chapter 7: Syrian Oil Production

U.S. soldiers continue to be put in harm's way to benefit Israel's energy security.

Why do we currently have soldiers in Syria? Could it be for the same oil interests that got us into Iraq in 2003? Our soldiers are protecting the largest illegal international oil smuggling operation in the world today.

When I saw President Trump announce in 2019 that we were keeping several hundred soldiers in Syria to protect the oil, my suspicions were aroused. The huge smirk on Senator Lindsey Graham's face almost guaranteed to me that the same interests that had pushed us into Iraq in 2003 were behind this latest decision to keep our soldiers in harm's way. The following quote in *Politico* further supports my suspicions of an oil agenda: "President Donald Trump declared… that the U.S. mission in Syria is focused solely on protecting oil fields, which appears to contradict the Pentagon's contention that fighting ISIS is the priority."[1]

Kurds' Oil

Close to 200 oil trucks a day crossed the border from Syria, delivering their oil to the Lanaz Refinery in Erbil, the largest city in Iraqi Kurdistan. The *Iraq Oil Report* identified 40,000 barrels a day being delivered from Syria to Lanaz Refinery in July 2022. That is about $4 million a day at international prices. Where is the money going? Has there been any accounting of this revenue made to the U.S. government? The photos of our troops in Syria consistently show them at oil locations or providing convoy protection for the tanker trucks carrying the oil. Are we reimbursed for this oil security from the oil revenues? I suspect not.

I asked my friends in Iraqi Kurdistan where they thought the Syrian oil was going. My sources advised me that the buyers of the Syrian crude oil in Iraqi Kurdistan had negotiated a highly discounted price from the Syrian Kurds. The Syrian Kurds do not have many customers for their production, so they are forced to accept the lower price. Much of the market value of those daily 40,000 barrels results in higher profits to the Iraqi Kurds.

This Syrian oil is owned by the sovereign government of Syria in Damascus. It is not the property of the Syrian Kurds who are pumping and trucking it out of Syria. Without the permission of Damascus, this operation is considered a smuggling operation by the international oil community. This situation is just like the oil operations previously conducted by the Iraqi Kurds before they negotiated a temporary agreement with the Iraqi federal government. The Iraqi Kurds delivered most of their oil to Israel, a country not recognized by Baghdad. As I wrote in Chapter 4, the Iraqi Kurds tried to deliver oil to the U.S. in 2014, but the Baghdad government successfully executed a seizure order through a U.S. judge in Galveston, Texas. One ship (*United Kalavryta*), however, was able to escape U.S. Customs in 2014 and eventually delivered its cargo to Israel.[2]

Another tanker (*Neverland*) shipped one million barrels of Kurdish oil to a refinery in Canada in mid-2017. The Federal Court of Canada ordered the seizure of the vessel if it entered Canadian territorial waters. But the ship turned off its radio beacon for several weeks, and four weeks later, it appeared in Malta, empty of its cargo.[3] The traders are still unapproachable today about where the oil was offloaded.

How does the Syrian oil benefit Israeli energy security interests? My contacts in Iraqi Kurdistan mentioned that the Iraqi Kurds were forced to decrease their oil exports to Ceyhan, Turkey, because their oil field production was decreasing. The Syrian oil being delivered to the Erbil refinery will enable the Iraqi Kurds to maintain higher export volumes to Ceyhan. The Kurdistan Regional Government (KRG) exported nearly 600,000 barrels a day at its peak to the traders.[4] Export volumes have recently decreased by as much as 30%.

The U.S. military is not the only agency supporting the smuggling operations in Syria. I received a phone call in 2019 from a retired U.S. Navy admiral who was running a defense contracting firm in the D.C. area. He was given my name as a Middle East oil expert who could help his company submit a successful bid for work in Syria. USAID, under the State Department, had announced a public offering to any competent contractor to bid on a contract for helping the Syrian Kurds increase their oil production.

My first comment to the CEO was that he should have his company lawyers review all the details of the contract thoroughly. If USAID was not offering legal protection to the successful bidder of the contract, then the risk would have to be absorbed by that company. I then explained to him that the Syrian oil operations were the largest international oil smuggling operation in the world, and that any company helping the Syrian Kurds could potentially be held liable in future litigation by the sovereign government of Syria. He thanked me for my advice, and I did not hear from him again.

International law prevents an occupying army from doing much with the natural resources of an occupied territory. We followed the rule of law for the oil sector in Iraq. Figure E, declassified in 2015, indicates the legal considerations that were followed in Iraq during the U.S. occupation.[5]

I was with a friend and recently retired Pentagon general officer in 2021, and I asked him why we continue to have U.S. troops in Syria. He insisted that the U.S. still needs to project its power in the Middle East. When I asked him whether oil has had anything to do with our decision, he responded that he did not know what I was talking about. I shared the full story of how vital Syrian oil was to the Iraqi Kurds and to the Israelis. I then highlighted that while we followed the rule of law in the oil sector in Iraq, I sensed that the rule of law was of no concern to us in Syria. My friend had no comment.

We currently have no legal right to be in Syria. Our Congress has not even declared war on Syria. Our soldiers are there without United Nations authorization and without the invitation of the sovereign government, providing security for the largest smuggling operation in the international oil industry. Do we no longer have any respect for the rule of law?

Natural Gas Discoveries in Eastern Mediterranean

Israel's long-term energy security solution is linked to the many newfound natural gas fields discovered in the Eastern Mediterranean since 2000. During the late 1990s, British Gas (BG) confirmed for the Palestinian Authority (PA) the presence of large quantities of gas offshore from Gaza. The PA quickly signed a contract with BG that would have given Palestine a meager 10% of the revenues, still worth billions of dollars. However, surrendering a massive gas bonanza to the Gazans was unthinkable to the Israeli government. Israel's prime minister Ehud Barak promptly canceled the BG deal.[6]

The *Oil & Gas Journal* (*OGJ*) has been reporting on several new gas discoveries in Israeli waters and along the maritime borders with Israel's neighbors. Since the early 2000s, the *OGJ* has verified that there is more than enough gas to provide for Israel's needs for several decades. The international oil industry is ready to assist in developing these new fields. A search of the *OGJ* archives yields more than 300 articles discussing the new

Levantine Basin discoveries, contract announcements, oil company accomplishments, Israeli government actions, etc. Most articles have been written by *OGJ* Exploration & Development reporters Alex Procyk and Tayvis Dunnahoe.[7] Chevron is the largest American oil company participating in developing the fields.[8]

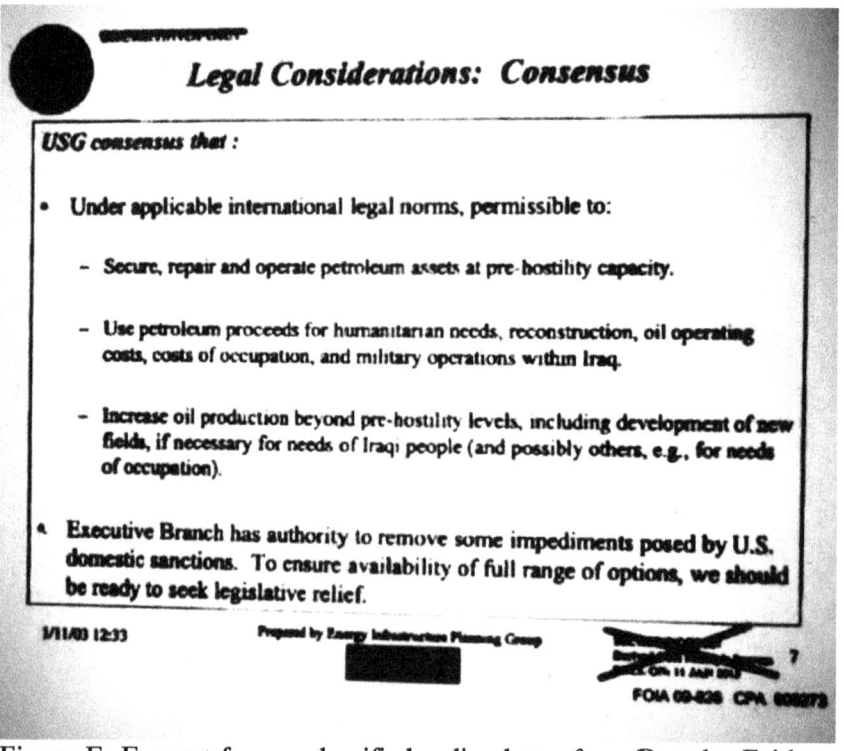

Figure E: Excerpt from a classified policy letter from Douglas Feith to Donald Rumsfeld dated January 16, 2003, declassified on May 19, 2015, detailing the U.S.'s legal considerations in the Iraq oil sector during occupation.

A very concise summary of the history of the gas discoveries, a map of the gas fields, political issues contributing to the delays, and potential size of the fields is found in chapter 8 of Charlotte Dennett's 2020 book *Follow the Pipelines*. It particularly highlights the 36 trillion cubic feet of natural gas in the Tamar and Leviathan gas fields that will offer "unprecedented energy security" for the country of Israel.[9]

This is an important point for oil policy in the region: as Israel increases its offshore gas production, Iraqi Kurdistan oil exports will become less important to Israeli energy security. However, oil traders such as Glencore and Trafigura will continue coveting the oil volumes they receive from Kurdistan to meet their profitability objectives.

Chapter 8. What Still Haunts Me

Power tends to corrupt, and
absolute power corrupts absolutely.

How did we allow the illegal 2003 invasion of Iraq to happen? Bush was highly motivated not only to topple Saddam Hussein but also to avenge the (alleged) assassination attempt on his father a decade earlier, but he also wanted to get re-elected to a second term in 2004.[1] To avoid becoming a one-termer like his father, Bush 43 provided significant clout to the Israel lobby in his first term, appointing their people to important positions at the Pentagon, in the Office of the Vice President, and on the NSC staff.[2] Much of what was included in "What Haunts Me," the final chapter of my 2017 book, still haunts me. Jay Garner told me to stop beating myself up about 2003, but that is easier said than done. I am still haunted. Therefore, in this final chapter of this new book, many items warrant mentioning again.

Neocons and War-Hawks

But how could removing Saddam best help the neoconservatives in their goal of assisting Israel during the Bush administration? Since the First Gulf War in 1991, Saddam had been contained by our military stationed in the Persian Gulf. Bush needed a team in his administration favorable to the Israel lobby and just as motivated as he was to remove Saddam. Where could he find such a team? The Project for the New American Century (PNAC) helped identify that team. PNAC, formed in 1997, became an important neoconservative voice in D.C. The signatories of the PNAC statement of principles on regime change in Iraq included neocons like Dick Cheney, Lewis Libby, Paul Wolfowitz, Zalmay Khalilzad, Peter Rodman, Richard Perle, and others, such as John Bolton.[3] All of these people would hold important positions in the Bush administration. None of them, except for John Bolton, ever wore the military uniform of the U.S. These people were among the "twenty-five people" without whom "the Iraq War would not have happened."[4]

An April 2003 *Guardian* article spoke of the history of key people in the Bush administration and their dream of getting Iraqi oil for Israel.* It quoted a former senior CIA official: "It has long been a dream of a powerful section of the people now driving this administration of President George W. Bush and the war in Iraq to safeguard Israel's energy supply as well as that of the United States."[5]

An effort was made by the U.S. government in 1983 to install a pipeline to move Iraqi crude oil to the port of Aqaba, adjacent to the southern end of the Israeli pipeline. Such volumes could have replaced Iranian crude oil on the secret Israeli pipeline. Donald Rumsfeld was CEO of G.D. Searle & Co. in the early 1980s. Rumsfeld was appointed by President Reagan as a presidential envoy to visit many Middle East countries. He even visited Saddam in Baghdad on December 20, 1983.

The Iran-Iraq War had taken a toll on Saddam's ability to export his oil. Iran had shut down Iraq's oil exports through the Basra ports. Syria, Iran's ally at the time, shut down Iraq's export pipeline through Syria. Saddam was totally dependent on exports through Turkey to the Mediterranean. One of Rumsfeld's agenda items was to help Saddam transport Iraq's crude through a different route. Rumsfeld proposed an export pipeline through Jordan to Aqaba, adjacent to the Israeli port of Eilat at the northeast end of the Gulf of Aqaba. Such a pipeline would have enabled Israel to access Iraqi crude oil for their secret pipeline rather than the Iranian crude oil that was under U.S. sanctions.[6]

An initiative to send a private citizen as a presidential envoy to the Middle East would have required significant coordination through the State Department. Paul Wolfowitz, Director of Policy Planning at the State Department in 1981–82, was moved to Assistant Secretary of State for East Asian and Pacific Affairs in 1983. Wolfowitz brought Lewis Libby, his former student and acolyte at Yale in the 1970s, into the State Department to work for him during this period.[7]

During the summer of 2003, when it became obvious to Chalabi and the neoconservatives that the Haifa pipeline plan was not feasible because the Iraqis would attack any pipeline sending oil to Israel, an alternative plan was executed. Chalabi's backup plan of ordering SOMO to sell oil to Glencore in October 2003 was the best he could deliver under the circumstances. In

---

* This connection is also discussed in Chapter 6 of *Iraq and the Politics of Oil*.

2014, the Kurds started delivering their exports to Israel, and that has proved to be much more profitable for Glencore than buying from SOMO. The supply arrangement with the Kurds has provided most of Israel's oil — and at discounted prices, too, like what they enjoyed from the Iranians prior to 1995.

Billions of U.S. dollars were at stake with a Haifa pipeline. The suggested pipeline size would have allowed transport of $30–50 billion per year of Iraqi oil through Israel. Was it possible that the neocons were motivated by the opportunity for financial gain with an Iraqi pipeline to Haifa? Would many of Friedman's "twenty-five people" get access to the economic gains from a pipeline to Haifa? I think that it is naïve to believe that greed did not play a role in their motivation to push the Iraq War. How else could the highly educated and brilliant neocons have fallen so easily for the lies of a known opportunist like Ahmed Chalabi? How were they so easily blinded to what was obvious to the CIA and State?

Chalabi for Prime Minister

First, senior political appointees at the Pentagon pushed for Ahmed Chalabi as prime minister. Secretary Rumsfeld may not have officially supported Chalabi for Iraq's first post-Saddam prime minister position, but there is no doubt that his subordinates were officially and unofficially supporting Chalabi. My Kuwait roommate in April 2003, Terry Sullivan, was a retired Navy SEAL. He and Dr. Harold Rhode had been sent to Kuwait by Doug Feith to link up with Chalabi's people. Michael Makovsky informed our prewar planning team in 2002 that Chalabi would be prime minister. Retired Lieutenant General Jay Garner was the first civilian head of Iraq. Jay related to me in 2016 that Doug Feith had met with him just before his departure for Kuwait. Feith suggested that Jay turn Iraq over to Chalabi shortly after Baghdad fell.* Feith became livid with Jay a few days later when Jay answered questions at a press conference before departing from the Pentagon for Kuwait. A particularly controversial answer to a question was Jay's claim that he and the U.S. government had not chosen Chalabi or any other specific Iraqi to lead Iraq after the fall of Saddam.[8]

It did not take long for Chalabi to walk away from his promise to the neocons of opening a pipeline to Haifa. Anyone with an understanding of

---

* This is also covered in Chapter 3 of *Iraq and the Politics of Oil*.

Iraqi society would know that Chalabi could not have delivered on his guarantee, but the neocons nevertheless had a strong motivation to believe his prewar promises. His duping of the neocons was uncovered in 2004.[9]

Chalabi did the only thing he could do in 2003 to provide oil secretly to Israel. He overruled the CPA directive of not selling oil to traders who had engaged in prewar corruption with Saddam's regime. He ordered the selling of Iraqi oil to Glencore (formerly Marc Rich & Co. AG).[10]

Chalabi tried to dupe the CIA and State Department into buying into his agenda for removing Saddam after 1991. Those agencies tried working with him, but after recognizing that he was not trustworthy, they eventually distanced themselves. Among his critics, Chalabi had a reputation for being unscrupulous. His supporters would not defend him. Chalabi found a more hospitable group in the powerful American neoconservatives with contacts within the Israel lobby. These gentlemen were all very intelligent and had experience within previous Republican administrations and in the Middle East. Chalabi had failed with CIA and State, but he had no problem convincing this group to support him.[11] Why was it so easy for him to deceive the neocons into supporting his Iraq War agenda? What was he selling to them that he did not sell to the CIA and State? Was greed part of the equation? If one believes Milton Friedman's theory that "societies run on greed," then greed for oil wealth and an oil agenda involving Israel was an easy sell.

During the late 1990s, Ahmed Chalabi promised the Israel lobby that he would provide Iraqi oil to Israel through the Haifa pipeline if the lobby would support him as the first prime minister post-Saddam. Chalabi delivered a rousing speech to JINSA in June 1997, during which he proposed an expansion of economic ties between Iraq and Israel by restoring the Kirkuk-Haifa pipeline.

K3 Destruction

The only infrastructure target deliberately destroyed during military operations was the K3 pump station,* which pumped crude oil from Anbar Province to ports in Syria prior to the invasion. While K3 was still burning from our attack, Secretary Rumsfeld stated to the press that the pump station had been destroyed to penalize Syria for enabling Saddam to bypass

---

* As discussed in Chapter 6 of *Iraq and the Politics of Oil*.

sanctions prior to the invasion.[12] Such an explanation was nonsense, considering how Jordan and Turkey had also assisted Saddam. The future government of Iraq was the true recipient of the punishment.

An action memo from Feith to Rumsfeld was issued on January 16, 2003 (declassified on May 19, 2015), with the subject: "Enabling Execution of Contingency Plans for Iraqi Oil Infrastructure." It requested that Rumsfeld direct the Army to be the Executive Agent for the program developed by the OSD-led interagency group dealing with Iraqi oil issues. All of these policies had been previously reviewed with President Bush's cabinet. One of the policies of the program specifically directed the military to "keep the [Iraq-Syria export pipeline] open if Syria cooperates on issues of interest to U.S./Coalition."[13] Paul Wolfowitz verbally reversed those written orders after they were issued, during a meeting with General Franks, shortly before the invasion. I was present during the meeting.

Our mission was to restore the oil sector to prewar levels, but Phil Carroll was ordered by the Pentagon civilians not to reconstruct the destroyed K3 pump station. A credible and logical explanation for the destruction of the K3 pump station was to remove any competition to future Haifa pipeline expansion. I was advised in 2015 by a former NSC member that Wolfowitz made decisions in 2002 and 2003 that were well above his pay grade. This was obviously one of them.

The following quotation in the *Guardian* shortly after the K3 destruction identified the motivation: "[The Haifa pipeline's] resurrection would transform economic power in the region, bringing revenue to the new U.S. dominated Iraq, cutting out Syria and solving Israel's energy crisis at a stroke."[14]

Makovsky's Israel Angle

Michael Makovsky contacted me from the Pentagon during the summer of 2003 trying to obtain a status report on the Haifa pipeline. This was my first exposure to the existence of the old Haifa pipeline. Makovsky characterized his inquiry to me as a top-priority project from the civilian leadership at the Pentagon. He hung up the phone on me when I realized that his inquiry was about a very sensitive pipeline to Israel and I requested that he submit his request in writing.* In Steve Coll's book *Private Empire: ExxonMobil and*

---

* This is covered in detail in Chapter 10 of *Iraq and the Politics of Oil*.

American Power (2013), Makovsky claimed that he was responding to an Israeli inquiry about the pipeline.[15] Why would Makovsky have been responding to an inquiry from the Israelis?

Michael Makovsky had linkages to the Israeli government. It was revealed to me in 2005 that Makovsky had left the U.S. for Israel in 1989 to join the Israeli foreign service.[16] In December 2005, I asked Makovsky about the reports of his time in the Israeli Defense Forces (IDF) and Israeli foreign service, and his response was that he would neither confirm nor deny these accusations.

Phil Carroll advised me in 2005 that Makovsky received his top-secret clearance only after Doug Feith waived the challenges of the Pentagon security group. Mike Mobbs, supervisor of our prewar interagency planning team, said that Makovsky was hired at the Pentagon in 2002 because of his AIPAC contacts. AIPAC was involved in the 2005 espionage scandal wherein Doug Feith's office was caught sharing classified documents with Israel.[17] This raised my suspicions that Makovsky was some type of an agent for the Israeli government. It also raised my suspicions about why an Israeli agent with little to no oil and gas experience was assigned as the Pentagon's Iraq oil contact. Makovsky arrived in 2002 like a new personal computer with a fresh install of an operating system, but rather than Microsoft Windows, it was Israel.

The Israeli infrastructure minister, Joseph Paritzky, publicly announced on March 31, 2003, his intention to reopen the Kirkuk-Haifa pipeline and receive Iraqi oil by the end of 2003. While U.S. troops were still fighting their way to Baghdad, Minister Paritzky was making public announcements of how Israel would benefit economically from the war. *Haaretz* quotes Paritzky as saying that a reopened Kirkuk-Haifa pipeline would cut Israel's energy bill drastically, by more than 25%. Israel, very dependent on imported oil to meet its energy needs, was getting most of its oil from Russia at the time, and Paritzky felt that they were paying too much of a premium for that oil.[18]

The Israeli government was described in *Haaretz* later in the summer of 2003 as having worked closely with sources inside the Pentagon. A telegram from the Pentagon in August to the Israeli government requested that Israel look into the possibility of pumping oil from Iraq to Haifa. According to *Haaretz*:

> The United States has asked Israel to check the possibility of pumping oil from Iraq to the oil refineries in Haifa. The request came in a

telegram last week from a senior Pentagon Official to a top Foreign Ministry official in Jerusalem.

The Prime Minister's Office, which views the pipeline to Haifa as a "bonus" the U.S. could give to Israel in return for its unequivocal support for the American-led campaign in Iraq, had asked the Americans for the official telegram.[19]

Marc Rich and Scooter Libby

There was a secret Israeli pipeline between the Red Sea and the Mediterranean. This pipeline through Israel carried Iranian crude oil and accounted for most of the wealth that Marc Rich accumulated. It flowed before and after the U.S. imposed sanctions on Iran. Rich and his lawyers kept the transactions confidential so that no one knew that Iranian crude oil was being delivered to Israel. It was a win-win for everyone involved: the Iranians had a customer for their crude oil in the Mediterranean, the Israelis received access to inexpensive Iranian crude oil for their refineries, and Marc Rich made billions of dollars.

The Internal Revenue Service investigated Rich's business during the 1980s for evading over $100 million in back taxes, and the FBI listed him as one of their top international fugitives. The title of the Interpol arrest warrant in 1985 was "Wanted International Criminal," and the charges included income tax evasion, racketeering, and trading with the enemy.[20] Late on January 20, 2001, his last day in office, President Clinton infamously signed a pardon for Rich and his partner, Pinkus Green.[21] Lewis Libby testified as a character witness on behalf of Marc Rich.[22]

Libby enjoyed a close working relationship with Paul Wolfowitz for more than 30 years, starting as a Yale political science student in a course taught by Wolfowitz in the early 1970s. Wolfowitz brought Libby into important jobs in the Reagan and Bush 41 administrations. Wolfowitz, Libby, and Zalmay Khalilzad were the primary writers of the 1992 Defense Planning Guidance (DPG). The DPG is widely regarded as an early formulation of the neoconservatives' post-Cold War agenda, laying out a series of economic and military objectives that were intended to ensure a U.S.-led unipolar global system.[23] Richard Perle and Dick Cheney were consultants during the drafting of that document. President George H.W. Bush, when he read the draft national security strategy based on the 1992 document, directed his staff to "send this back to the crazies in the basement of the Pentagon…"

Conclusion

The costs of invading Iraq were huge. Some of those costs included: 4,489 American sons and daughters killed, 32,223 troops injured (not including those who have suffered from PTSD), more than 250,000 Iraqis killed, 150 reporters killed, 2.8 million people who remain either internally displaced or have fled Iraq, and a financial cost of more than $2 trillion.[24] Did my country incur all these costs because of a few dozen neocons? Did a network of highly educated yet conflicted U.S. government senior political appointees develop a plan to enrich themselves and their colleagues? Did they bring in staffers with a similar ideology to help execute such a plan? Some staffers, like Michael Makovsky, arrived in 2002 with direct ties to the Israeli government.

Furthermore, neocons from this very group are still delivering speeches, promoting the next Middle East war, lecturing at universities, appearing as guests on television political shows, and receiving honoraria. They feel no shame, no remorse. What haunts me is that I was intentionally misled into believing that confiscating WMDs was a noble reason for war, when it was really just an excuse for their war. I was an unknowing participant in their oil scheme.

What haunts me are the grieving parents, spouses, and children of those soldiers who made the ultimate sacrifice. Or the caregivers of those disabled soldiers who can no longer care for themselves. It is essential that the survivors understand the noble cause of the military profession. It is a principled cause of duty to country and a love of country. There is something exclusive about raising your right hand and swearing to defend the Constitution of our country against her enemies. This is a noble cause. It is a cause that brings patriotic Americans together on a military team. Being part of a military team engaged in battle is an event that few experience. The brotherhood of those team members is extremely close. Studies have shown that soldiers are not motivated during battle because of their president, or congressmen, or other political leaders. Instead, soldiers fight for each other.[25] It is the honor of their profession to die for their fellow soldiers. It is a creed among brothers and sisters. It is their creed of duty, honor, and country.

America must be vigilant in the future. We cannot allow another group such as the neocons to push our country into another costly and unnecessary war. Unfortunately, I am not optimistic. The Bush administration was full of conflicted civilians in key senior policy positions. Any concerns about their security clearances were waived. The results of their conflicts of interest were devastating. No one has been held accountable, and no changes to the laws or procedures have since been made to prevent this type of abuse, so it has continued into the current administration. It is only a matter of time before another grossly conflicted group pushes us into another war that is not in our national interests. The neocons wanted our military to topple Syria and Iran back in 2003 after Iraq, and only the still-naïve would believe that their objectives have changed.

Power tends to corrupt, and absolute power corrupts absolutely. That is why I am not optimistic about the U.S. *not* repeating the Iraq blunder. It will happen again. The power base of the neocons is the powerful Israel lobby, primarily AIPAC. The Israel lobby has controlled U.S. Middle East policy for decades. A former 30-year U.S. foreign service officer whom Ambassador Bremer brought to Baghdad in 2003 was his trusted former State Department colleague. He became one of Bremer's senior communications advisors. He first enlightened me in August 2003. After I asked him several questions about our Middle East policies, he said that I needed to understand that the U.S. does nothing in the Middle East — and he emphasized *nothing* — without the direction or approval of the Israel lobby. I did not believe him initially, but everything I have read or experienced since has supported this guidance. I now know that anyone who tells me anything different is lying to me.

Many others with Middle East diplomatic experience have reinforced his advice to me. The last U.S. president who had the courage to defy the lobby openly was Ronald Reagan in 1986, concerning the sale of AWACS (Airborne Warning and Control System) planes to Saudi Arabia. The result was that AIPAC took that setback as a defeat and transformed itself with the financial help of some very wealthy donors. Their budget climbed eightfold, and membership fivefold, in a few years. It gained greater power than ever before and was described as a "superlobby" by the Pulitzer Prize-winning author Hedrick Smith in his 1988 book *The Power Game*.[26] Absolute power was achieved.

Other books further analyze the power of the lobby. One of the most notable has been *The Israel Lobby and U.S. Foreign Policy* (2008), written by two prominent professors, John Mearsheimer of the University of Chicago and Stephen Walt of Harvard.[27] There has been heavy criticism of Walt and

Mearsheimer, most of it heavily funded and orchestrated by AIPAC, no doubt. When President Obama was able — barely able — to push through the Joint Comprehensive Plan of Action (JCPOA) with Iran in his second term of office, it was a second resounding defeat for AIPAC, which lobbied heavily against the nuclear agreement. President Trump was then elected and withdrew the U.S. from the agreement. So, in the end, AIPAC triumphed. When will the American people awaken to this dangerous and powerful foreign influence on our national security decision-making?

The very first sentence of this book stated that it is important that my story be told. I hope that it has enlightened readers on the importance of fully disclosing what really happened in Iraq. To quote Winston Churchill: "You can always rely on America to do the right thing — once it has exhausted the alternatives." Americans are good people, and courageous Americans will do the right thing, if they know the truth. That is my hope!

Appendix A:
Unclassified Email – DEPSECDEF Inquiry of Oil for Israel

Below is a recreation of the email chain referenced in Chapter 4. For a full list of recipients, visit: archive.org/details/re-swd-rio-sitrep-27-august.

From: Clark, Mitchell L. (GS-11) and Michelle
Sent: August 27, 2003, 12:37 PM
To: (See full list at archive link.)
Subj.: SWD RIO SITREP 27 August

v/r

Michelle L. Clark

FEST-M RIO
U.S. Army Corps of Engineers
OCPA-Oil-TF-RIO

From: Jewell, James S.
Sent: August 27, 2003, 9:44 PM
To: Clark, Mitchell L. (GS-11) and Michelle; et al. (incl. Vogler, Gary)
Subj.: RE: SWD RIO SITREP 27 August

Does anybody know if we are selling Iraqi oil to Israel?

From: Vogler, Gary
Sent: August 27, 2003, 11:15 PM
To: Jewell, James S.; et al.
Subj.: RE: SWD RIO SITREP 27 August

Jim,

We receive information from SOMO indicating the name of the company purchasing the oil. We get no visibility on the destination. Is this an issue?

Gary

From: Jewell, James S.
Sent: August 30, 2003, 2:09 PM
To: Vogler, Gary
Subj: RE: SWD RIO SITREP 27 August

Gary,

Don't know. Probably be hard to get final destination from the purchaser. Although, I guess it would not hurt to ask. This has been of interest in the past, and it might be good to know if there were any potentially embarrassing destinations. Depsecdef asked the question, but I think it had more to do with comments in the press regarding pipelines after further investigation. Thanks for the response.

Regards,
Jim J.

Appendix B:
"A Whole New Ballgame Overseas"

by Jack Herman
St. Louis Post-Dispatch, February 20, 1989

As a high-level aide to John C. Danforth, Michael Makovsky discovered early that he shared at least two interests with the U.S. senator from Missouri.

"Danforth is an avid Cardinals fan and genuinely cares about Israel," Makovsky said, and those are "subjects that I'm really interested in."

When he leaves soon for Israel, he's "really going to miss listening to their games and reading about the Cardinals."

Makovsky, a *Post-Dispatch* scholar-athlete eight years ago, now hopes to launch a career in Israel's foreign service. He was born in St. Louis, is an alumnus of Block Yeshiva High School here and graduated with honors from the University of Chicago, where he majored in history of American foreign affairs, with a minor in British history.

Apparently the 25-year-old St. Louisan found the road to Washington a natural progression. He worked as a freelance researcher there for Republicans on the Senate Foreign Relations Committee before joining Danforth's staff in the summer of 1987.

After six months as a legislative correspondent for Danforth, Makovsky was offered a full-time job as Danforth's assistant press secretary.

Danforth found Makovsky useful in pressing the Soviet Union on human rights.

"I wrote a letter to Gorbachev," Makovsky recalled, "urging the release of Joseph Begun and Gregory Rosenstein. Begun was a leader of the original 'refuseniks.'

"In fact, I put through a call to Begun in Moscow and translated, in Hebrew and English, a conversation between him and Danforth."

Two months later, Begun was allowed to immigrate to Israel.

Makovsky feels strongly about moving to Israel, where he'll have to serve a stretch in the army before entering the country's foreign service.

Having attended a Jerusalem high school for a semester in 1978, he began focusing on a possible move to Israel three years ago.

After his graduation from the University of Chicago, he worked for then-Prime Minister Shimon Peres.

"I would write weekly news summaries on what United States and British newspapers were saying about Israel in general and Peres in particular," Makovsky noted.

"Peres, of all Israeli officials, is sensitive about his image and foreign perception of his country."

Peres now is finance minister under the new coalition government.

Makovsky also wrote media analyses about the Pollard spy case, which strained relations between the United States and Israel two years ago. Pollard currently is serving a prison term in the federal penitentiary in Marion, Ill.

Makovsky's immediate superior during this period was Uri Savir, now consul general for Israel in New York. Savir served as an intermediary between Makovsky and Peres and, last November, encouraged Makovsky to enter the Israeli foreign service.

Makovsky's father, Donald I. Makovsky, is executive director of Block Yeshiva High, which graduated its first class eight years ago. The younger Makovsky was a member of that class. He excelled in track, basketball and cross country.

His father and grandfather are also fervent fans of the Cardinals. His grandfather, Jack Makovsky, "took me to my first game when I was only 5," Michael Makovsky recalled, "and I've been hooked ever since."

Actually, the Makovsky family can be said to be a farm system for Israel. Another son, David, 28, currently is working on his master's degree in journalism at Harvard.

"David and I have similar interests," Makovsky said. "He's a journalist, and I hope to be the diplomat in the family. When he leaves Harvard, we'll meet in Jerusalem. He will be a political correspondent for the *Jerusalem Post* as well as Israeli correspondent for U.S. News and World Report."

Makovsky doesn't delude himself about leaving the United States and living in Israel.

"I recognize the difficulties involved," he said. "This is a democratic, fun country, and living in suburbia makes it all the more difficult to leave.

"But I have strong feelings about helping to build a Jewish state. There's an excitement about living in a young country that's only 41 years old, a country with problems on its borders. It's like returning to your roots."

Under Israel's Law of Return, Makovsky immediately will enjoy dual citizenship. He doesn't plan to renounce his U.S. citizenship.

Makovsky speaks fondly of his association with Danforth. "We both wanted to read the *Post*'s sports section about what the Cardinals did the night before," he said. "If I got the paper first, I'd copy the sports section for him. And at night, I'd listen to KMOX and the play-by-play of Cardinal games."

His experience and background with both the Israeli and U.S. governments, Makovsky said, should operate in his favor. The foreign service exam is a seven-month process, and he concedes that his move to Israel "is not practical and was not an easy decision."

"But I'm looking forward to returning here in the fall for the playoffs and World Series," he said.

Endnotes

Chapter 1

[1] Tim Adams, "Iraq war whistleblower Katherine Gun: 'Truth always matters,'" *Guardian*, September 22, 2019, theguardian.com/film/2019/sep/22/katharine-gun-whistleblower-iraq-official-secrets-film-keira-knightley.

[2] "Did the First President Bush Lose His Job to the Israel Lobby?" *Observer*, July 17, 2006. observer.com/2006/07/did-the-first-president-bush-lose-his-job-to-the-israel-lobby.

[3] John Nixon, *Debriefing the President: The Interrogation of Saddam Hussein* (New York: Penguin Press, 2016).

[4] John King, "Bush calls Saddam 'the guy who tried to kill my dad,'" CNN, September 27, 2002, edition.cnn.com/2002/ALLPOLITICS/09/27/bush.war.talk.

[5] Robert Draper, *To Start a War: How the Bush Administration Took America into Iraq* (New York: Penguin Press, 2020).

[6] Seymour Hersh, "A Case Not Closed," *New Yorker*, November 1, 1993, newyorker.com/magazine/1993/11/01/a-case-not-closed.

[7] Ari Shavit, "White Man's Burden," *Haaretz*, April 3, 2003, haaretz.com/2003-04-03/ty-article/white-mans-burden/0000017f-e398-d804-ad7f-f3fa5d520000.

[8] Richard Perle, James Colbert, Charles Fairbanks, Jr., Douglas Feith, Robert Loewenberg, Jonathan Torop, David Wurmser, Meyrav Wurmser, "A Clean Break: A New Strategy for Securing the Realm." Study Group on a New Israeli Strategy Toward 2000, Institute for Advanced Strategic and Political Studies, 1996. mafhoum.com/press7/realm2.htm.

[9] Michael Marshall, "Foreign Policy Experts Assess War on Terrorism, Saber Rattling Toward Iraq," University of Virginia School of Law, September 11, 2002, law.virginia.edu/news/2002_fall/terrorismforum.htm.

[10] "John Mearsheimer and Stephen Walt: 'The Israel Lobby,'" *London Review of Books* 28, No. 6, March 23, 2006, lrb.co.uk/the-paper/v28/n06/john-mearsheimer/the-israel-lobby.

[11] "Generals and Admirals Program to Israel." Accessed December 7, 2022, jinsa.org/jinsa_program/general-admirals-trip-israel.

Chapter 2

[1] Stephen Scheer, "Netanyahu says Iraq-Israel oil line is no Pipe-Dream," *Haaretz*, June 20, 2003, haaretz.com/2003-06-20/ty-article/netanyahu-says-

iraq-israel-oil-line-not-pipe-dream/0000017f-ef24-ddba-a37f-ef6ebbb70000.

2. Dr. Ferruh Demirmen, "Oil in Iraq: The Byzantine Beginnings," *Global Policy*, April 25, 2003, archive.globalpolicy.org/security/oil/2003/0425byzantine.htm.

3. Gary Vogler, *Iraq and the Politics of Oil: An Insider's Perspective* (Lawrence: University Press of Kansas, 2017), 36.

4. Ibid.

5. David L. Phillips, *Losing Iraq: Inside the Postwar Reconstruction Fiasco* (New York: Basic Books, 2006), 71.

6. Daniel Amman, *The King of Oil: The Secret Lives of Marc Rich* (New York: St. Martin's Press, 2009), 64–65.

7. Ari Shavit, "White Man's Burden," *Haaretz*, April 3, 2003, haaretz.com/2003-04-03/ty-article/white-mans-burden/0000017f-e398-d804-ad7f-f3fa5d520000.

8. Akiva Eldar, "Infrastructure Minister Paritsky Dreams of Iraqi Oil Flowing to Haifa," *Haaretz*, March 31, 2003, haaretz.com/2003-03-31/ty-article/infrastructure-minister-paritzky-dreams-of-iraqi-oil-flowing-to-haifa/0000017f-e2ca-d7b2-a77f-e3cf5dd90000.

9. Stephen Scheer, "Netanyahu says Iraq-Israel oil line is no Pipe-Dream," *Haaretz*, June 20, 2003, haaretz.com/2003-06-20/ty-article/netanyahu-says-iraq-israel-oil-line-not-pipe-dream/0000017f-ef24-ddba-a37f-ef6ebbb70000.

10. John Dizard, "How Ahmed Chalabi Conned the Neocons," *Salon*, May 4, 2004, salon.com/2004/05/04/chalabi_4.

11. Alan Weisman, *Prince of Darkness: Richard Perle: The Kingdom, the Power & the End of Empire in America* (New York: Union Square Press, 2007).

12. James Mann, *Rise of the Vulcans: The History of Bush's War Cabinet* (New York: Viking Press, 2004), 23.

13. Ibid., 367.

14. Thomas E. Ricks, *Fiasco: The American Military Adventure in Iraq, 2003 to 2005* (New York: Penguin Press, 2006), 78.

15. Salim Muwakkil, "For Israel's Sake," *In These Times*, March 13, 2007, inthesetimes.com/article/for-israel-sake.

16. Gary Vogler, *Lessons Learned – The Iraq Energy Sector* (Fairfax: Howitzer Consulting, 2016).

17. Gary Vogler, *Iraq and the Politics of Oil: An Insider's Perspective* (Lawrence: University Press of Kansas, 2017), 59–62.

18. "News Release — IMMEDIATE RELEASE No. 201-03." Office of the Assistant Secretary of Defense (Public Affairs), April 4, 2003. arlingtoncemetery.net/rbrippetoe.htm.

19 W. Patrick Lang. "Drinking the Kool-Aid." *Middle East Policy* 11, No. 2 (Summer 2004), 8.
20 Daniel Amman, *The King of Oil: The Secret Lives of Marc Rich* (New York: St. Martin's Press, 2009), 64.
21 Javier Blas and Jack Farchy, *The World for Sale: Money, Power, and the Traders Who Barter the Earth's Resources* (New York: Oxford University Press, 2022), 95–99.
22 Daniel Amman, *The King of Oil: The Secret Lives of Marc Rich* (New York: St. Martin's Press, 2009), 65.
23 Todd Garvey, "A Constitutional Anomaly: Safeguarding Confidential National Security Information Within the Enigma That Is the American Vice Presidency," *William & Mary Bill of Rights Journal* 17, No. 565 (2008).

Chapter 3

1 Ed Vuillamy, "Israel Seeks pipeline for Iraqi Oil," *Guardian*, April 19, 2003, theguardian.com/world/2003/apr/20/israelandthepalestinians.oil.
2 Gordon Rudd, *Reconstructing Iraq: Regime Change, Jay Garner, and the ORHA Story* (Lawrence: University Press of Kansas, 2011), 148.
3 Michael Gordon, "The Struggle for Iraq: Reconstruction; 101st Airborne Scores Success in Northern Iraq," *New York Times*, September 4, 2003, nytimes.com/2003/09/04/world/struggle-for-iraq-reconstruction-101st-airborne-scores-success-northern-iraq.html.
4 Hugh Pope, "Iraq Adds Complexity to U.S., Syria Relations," *Wall Street Journal*, October 20, 2003, ia902602.us.archive.org/31/items/iraq-adds-complexity-to-u.-s.-syria-relations-wsj/Iraq%20Adds%20Complexity%20To%20U.S.%2C%20Syria%20Relations%20-%20WSJ.pdf.
5 Ibid.
6 "JINSA IRS Form 990 for tax year 2019, Michael Makovsky, principal officer." Internal Revenue Service. Accessed June 2023, apps.irs.gov/pub/epostcard/cor/521233683_201912_990_2021030117771378.pdf.
7 Grant F. Smith, *Big Israel: How Israel's Lobby Moves America* (Washington, D.C.: Institute for Research, Middle East Policy, 2016), 10.
8 Kevin Ross and Gary Vogler, "Iraqis Mending Own Pipelines," *Oil & Gas Journal*, Feb 16, 2009, ia902709.us.archive.org/14/items/iraqis-mending-own-pipelineshttpswww.ogj.comprintcontent-17221706/Iraqis%20Mending%20Own%20Pipelineshttps%3A%3Awww.ogj.com%3Aprint%3Acontent%3A17221706.pdf.
9 Iraq Coalition Casualty Count, "Iraq Coalition Casualties: Fatalities by Year and Month." Accessed March 2016, icasualties.org/App/Fatalities.

¹⁰ "Wolfowitz Underestimates War Deaths," United Press International, April 29, 2004, upi.com/Top_News/2004/04/29/Wolfowitz-underestimates-war-deaths/17841083268328.

Chapter 4

¹ "Subj: SWD RIO Sitrep 27 August," Defense Intelligence Agency, unclassified email dated August 27, 2003. archive.org/details/re-swd-rio-sitrep-27-august. See Appendix A.

² Daniel Amman, *The King of Oil: The Secret Lives of Marc Rich* (New York: St. Martin's Press, 2009), 99.

³ Ibid., 103.

⁴ Javier Blas and Jack Farchy, *The World for Sale: Money, Power, and the Traders Who Barter the Earth's Resources* (New York: Oxford University Press, 2022), 202.

⁵ Ibid., 282.

⁶ "U.S. Disturbed by Iraqi Law Criminalizing Israel Ties," *Al-Monitor*, May 27, 2022, al-monitor.com/originals/2022/05/us-disturbed-iraqi-law-criminalizing-israel-ties.

⁷ Ben Van Heuvelen and Ben Lando, "UPDATE: Iraqi Supreme Court strikes down KRG oil sector independence," *Iraq Oil Report*, Feb 15, 2022, archive.org/details/update-iraqi-supreme-court-strikes-down-krg-oil-sector-independence-iraq-oil-report/page/n1/mode/2up.

⁸ Lizzie Porter and *Iraq Oil Report* Staff, "KRG Shifts Legal Tactics as Baghdad Court Delays Further Rulings," August 2, 2022, archive.org/details/krg-shifts-legal-tactics-as-baghdad-court-delays-further-rulings-iraq-oil-report/page/n3/mode/2up.

⁹ "Amos J. Hochstein." U.S. Department of State. Accessed December 26, 2023. state.gov/biographies/amos-j-hochstein.

¹⁰ Tal Schneider, "Four years after leaving post, will Dan Shapiro return as old-new U.S. ambassador?" *Times of Israel*, January 20, 2021, timesofisrael.com/four-years-after-leaving-post-will-dan-shapiro-return-as-old-new-us-ambassador.

¹¹ Jack Herman, "A Whole New Ballgame Overseas," *St. Louis Post-Dispatch*, February 20, 1989, ia902604.us.archive.org/11/items/makovsky-stltoday-sports/makovsky%20STLtoday%20-%20Sports%20-.pdf. See Appendix B.

¹² James Clapper, "Security Executive Agent Directive 4: National Security Adjudicative Guidelines," December 10, 2016, dni.gov/files/NCSC/documents/Regulations/SEAD-4-Adjudicative-Guidelines-U.pdf.

Chapter 5

¹ "Energy Security – About – IEA." International Energy Agency. Accessed May 2023, iea.org/about/energy-security.

² Daniel Amman, *The King of Oil: The Secret Lives of Marc Rich* (New York: St. Martin's Press, 2009), 64.

³ Ibid., 102.

⁴ Gary Vogler, "Oil Pipelines played role in U.S. invasion of Iraq," *Oil & Gas Journal*, December 3, 2018, web.archive.org/web/20230605210058/www.ogj.com/home/article/17232358/oil-pipelines-played-role-in-us-invasion-of-iraq.

⁵ Javier Blas and Jack Farchy, *The World for Sale: Money, Power, and the Traders Who Barter the Earth's Resources* (New York: Oxford University Press, 2022), 50–51.

⁶ Daniel Amman, *The King of Oil: The Secret Lives of Marc Rich* (New York: St. Martin's Press, 2009), 103.

⁷ Ibid.

⁸ Ibid., 183.

⁹ Ibid., 227.

¹⁰ Ibid., 225–229.

¹¹ Ibid., 242.

¹² Akiva Eldar, "Infrastructure Minister Paritsky Dreams of Iraqi Oil Flowing to Haifa," *Haaretz*, March 31, 2003, haaretz.com/2003-03-31/ty-article/infrastructure-minister-paritzky-dreams-of-iraqi-oil-flowing-to-haifa/0000017f-e2ca-d7b2-a77f-e3cf5dd90000.

¹³ Eytan Avriel, "Richest Israelis Are Getting Richer, and Other Takeaways from Haaretz's 2019 Rich List," *Haaretz*, June 19, 2019, haaretz.com/israel-news/business/2019-06-19/ty-article/.premium/richest-israelis-getting-richer-and-other-takeaways-from-haaretzs-2019-rich-list/0000017f-e558-df2c-a1ff-ff595dfb0000.

¹⁴ "Forbes Publishes 34th Annual List of Global Billionaires," *Forbes*, April 7, 2020, forbes.com/sites/forbespr/2020/04/07/forbes-publishes-34th-annual-list-of-global-billionaires/?sh=6f6e84f03edf.

¹⁵ "U.S. Imports from Iraq of Crude Oil and Products from 1996 to 2022." U.S. Energy Information Administration. Accessed May 2023, ia902703.us.archive.org/14/items/u.-s.-imports-from-iraq-of-crude-oil-and-petroleum-products-thousand-barrels/U.S.%20Imports%20from%20Iraq%20of%20Crude%20Oil%20and%20Petroleum%20Products%20%28Thousand%20Barrels%29.pdf.

Chapter 6

¹ "'Outing' the Neocons: The Office of Special Plans," The Daily Kos, March 12, 2005, dailykos.com/stories/2005/3/12/98993/-.

² Jack Herman, "A Whole New Ballgame Overseas," *St. Louis Post-Dispatch*, February 20, 1989, ia902604.us.archive.org/11/items/makovsky-stltoday-sports/makovsky%20STLtoday%20-%20Sports%20-.pdf. See Appendix B.

3 Steve Coll, *Private Empire: ExxonMobil and American Power* (New York: Penguin Press, 2013).

4 "Generals and Admirals Program to Israel." Accessed December 7, 2022, jinsa.org/jinsa_program/general-admirals-trip-israel.

Chapter 7

1 David Brown, "Trump says U.S. left troops in Syria 'only for the oil,' appearing to contradict Pentagon," *Politico*, November 13, 2019, politico.com/news/2019/11/13/trump-troops-syria-oil-pentagon-070567.

2 Matthew Philips, "A Mysterious Oil Tanker Might Hold the Key to Kurdish Independence," *Bloomberg Businessweek*, October 23, 2014, web.archive.org/web/20220703213732/https://www.bloomberg.com/news/articles/2014-10-23/iraqi-kurds-seek-independence-through-shady-oil-sales.

3 Javier Blas and Jack Farchy, *The World for Sale: Money, Power, and the Traders Who Barter the Earth's Resources* (New York: Oxford University Press, 2022), 286.

4 Ibid., 287.

5 Gary Vogler, *Lessons Learned – The Iraq Energy Sector* (Fairfax: Howitzer Consulting, 2016).

6 Charlotte Dennett, *Follow the Pipelines: Uncovering the Mystery of a Lost Spy and the Deadly Politics of the Great Game for Oil* (London: Chelsea Green Publishing, 2020), 207.

7 Tayvis Dunnahoe, "Israeli outreach aims to further exploration, development," *Oil & Gas Journal*, May 7, 2018, ia902709.us.archive.org/14/items/israeli-outreach-aims-to-further-gas-developmenthttpswww.ogj.comprintcontent-17232681/Israeli%20outreach%20aims%20to%20further%20gas%20development https%3A%3Awww.ogj.com%3Aprint%3Acontent%3A17232681.pdf.

8 Alex Procyk, "Chevron let pipeline contract for Tamar field expansion," *Oil & Gas Journal*, January 27, 2023, ogj.com/exploration-development/article/14288850/chevron-lets-pipeline-contract-for-tamar-field-expansion.

9 Charlotte Dennett, *Follow the Pipelines: Uncovering the Mystery of a Lost Spy and the Deadly Politics of the Great Game for Oil* (London: Chelsea Green Publishing, 2020), 211.

Chapter 8

1 Scott Horton, *Enough Already: Time to End the War on Terrorism* (Austin: Libertarian Institute, 2021), 65.

2 Ibid.

3 George Packer, "PNAC and Iraq," *New Yorker*, March 29, 2009, newyorker.com/news/george-packer/pnac-and-iraq.

4 Ari Shavit, "White Man's Burden," *Haaretz*, April 3, 2003, haaretz.com/2003-04-03/ty-article/white-mans-burden/0000017f-e398-d804-ad7f-f3fa5d520000.

5 Ed Vuillamy, "Israel Seeks pipeline for Iraqi Oil," *Guardian*, April 19, 2003, theguardian.com/world/2003/apr/20/israelandthepalestinians.oil.

6 Joyce Battle (ed.). "The National Security Archive No. 82," February 25, 2003, nsarchive2.gwu.edu/NSAEBB/NSAEBB82.

7 Bill Christison and Kathleen Christison, "The Bush Neocons and Israel," *CounterPunch*, September 6, 2004, counterpunch.org/2004/09/06/the-bush-neocons-and-israel.

8 Gordon Rudd, *Reconstructing Iraq: Regime Change, Jay Garner, and the ORHA Story* (Lawrence: University Press of Kansas, 2011), 142–3.

9 John Dizard, "How Ahmed Chalabi Conned the Neocons," *Salon*, May 4, 2004, salon.com/2004/05/04/chalabi_4.

10 Daniel Amman, *The King of Oil: The Secret Lives of Marc Rich* (New York: St. Martin's Press, 2009), 152.

11 Jane Mayer, "The Manipulator," *New Yorker*, May 30, 2004, newyorker.com/magazine/2004/06/07/the-manipulator.

12 David Sanger and Thom Shanker, "A Nation at War: White House; Bush Says Regime in Iraq is No More; Syria is Penalized," *New York Times*, April 16, 2003, nytimes.com/2003/04/16/world/nation-war-white-house-bush-says-regime-iraq-no-more-syria-penalized.html.

13 Gary Vogler, *Lessons Learned – The Iraq Energy Sector* (Fairfax: Howitzer Consulting, 2016).

14 Ed Vuillamy, "Israel Seeks pipeline for Iraqi Oil," *Guardian*, April 19, 2003, theguardian.com/world/2003/apr/20/israelandthepalestinians.oil.

15 Steve Coll, *Private Empire: ExxonMobil and American Power* (New York: Penguin Press, 2013), 221.

16 Jack Herman, "A Whole New Ballgame Overseas," *St. Louis Post-Dispatch*, February 20, 1989, ia902604.us.archive.org/11/items/makovsky-stltoday-sports/makovsky%20STLtoday%20-%20Sports%20-.pdf. See Appendix B.

17 Dan Eggen and Jerry Markon, "2 Senior AIPAC Employees Ousted," *Washington Post*, April 21, 2005, washingtonpost.com/archive/politics/2005/04/21/2-senior-aipac-employees-ousted/3bb1bafd-af60-4fbf-81a1-03cad44a0ca8.

18 Ed Vuillamy, "Israel Seeks pipeline for Iraqi Oil," *Guardian*, April 19, 2003, theguardian.com/world/2003/apr/20/israelandthepalestinians.oil.

19 Amiram Cohen, "U.S. Checking Possibility of Pumping Oil from Northern Iraq to Haifa, via Jordan," *Haaretz*, August 25, 2003, haaretz.com/2003-08-

25/ty-article/u-s-checking-possibility-of-pumping-oil-from-northern-iraq-to-haifa-via-jordan/0000017f-e93b-dea7-adff-f9fb42670000.

[20] Daniel Amman, *The King of Oil: The Secret Lives of Marc Rich* (New York: St. Martin's Press, 2009), 152.

[21] Ibid., 242.

[22] Ibid., 137.

[23] "1992 Draft Defense Planning Guidance," *Militarist Monitor*, February 20, 2020, militarist-monitor.org/profile/1992_draft_defense_planning_guidance.

[24] Michael B. Kelley and Geoffrey Ingersoll, "BY THE NUMBERS: The Staggering Cost of the Iraq War," *Business Insider*, June 20, 2014, businessinsider.com/iraq-war-facts-numbers-stats-total-2013-3.

[25] Robert J. Rielly. "Confronting the Tiger: Small Unit Cohesion in Battle." *Military Review* (November 2000), scribd.com/document/126229307/Confronting-the-Tiger-Small-Unit-Cohesion-in-Battle-Robert-J-Reilly-91637655-Small-Unit-Leadership.

[26] Hedrick Smith, *The Power Game: How Washington Works* (New York: Random House, 1988), 216–222.

[27] John J. Mearsheimer and Stephen M. Walt, *The Israel Lobby and U.S. Foreign Policy* (New York: Farrar, Straus and Giroux, 2008).

Acknowledgments

My 2017 book, *Iraq and the Politics of Oil*, was dedicated to my wife. We have been a great team for a half-century. Her support during my many trips to the Middle East was huge. The visits home during my 75 months of deployment always ended with very difficult farewells. I could not have spent all those months of deployment without her moral and emotional support. She is my primary editor and reviews everything I write. Her help with this book has been invaluable.

Several West Point classmates and friends have helped with my writing. Tom and Georgia Mason were extremely helpful to me with this book. Their counsel was always sought, welcomed, and valued during the process of writing the manuscript.

Other West Point graduates who reviewed and commented on this book included the Honorable Allen B. Clark, MG Dan Wright, Professor Emeritus Leslie R. Alm, Colonel John Scanlan, and Richard Dickinson. A special thanks to LTC Courtney Rittgers, my West Point Tactical Officer in 1973 who taught us to "never accept the unacceptable."

Others provided valuable comments. They included retired Professor (Colonel) Lawrence B. Wilkerson, Donald T. Phillips, and former CIA analysts John Nixon and Larry Johnson. Retired Navy Captain Kevin Ross, who worked with me in Baghdad for several years, also provided valuable input.

The maps were made by Joe LeMonnier of MapArtist.com, who also created the maps for my first book. Many thanks to Andrew Zehnder as well for designing and producing the cover art.

I would like to thank Scott Horton and the Libertarian Institute for publishing this book. Many thanks to Mike Dworski for his assistance in preparing this book for publication, and also to Ben Parker, who provided a final editing of this book.

Also by Gary Vogler

"Oil pipelines played role in U.S. invasion of Iraq"
Oil & Gas Journal, December 2018

Iraq and the Politics of Oil: An Insider's Perspective
University Press of Kansas, 2017

Lessons Learned – The Iraq Energy Sector
Amazon Kindle, 2016

"Iraq Crude Oil Export Expansion Heightens Country's Security"
Oil & Gas Journal, May 2012

"Iraqis Mending Own Pipelines"
Oil & Gas Journal, February 2009

The Libertarian Institute

Check out the Libertarian Institute at LibertarianInstitute.org. It's Scott Horton, Sheldon Richman, Laurie Calhoun, James Bovard, Kyle Anzalone, Keith Knight and the best libertarian writers and podcast hosts on the Internet. We are a 501(c)(3) tax-exempt charitable organization. EIN 83-2869616.

Help support our efforts — including our project to purchase wholesale copies of this book to send to important congressmen and women, antiwar groups and influential people in the media. We don't have a big marketing department to push this effort. We need your help to do it. And thank you.

LibertarianInstitute.org/donate or
The Libertarian Institute
612 W. 34th St.
Austin, TX 78705

Check out all of our other great books at LibertarianInstitute.org/books:
Hotter Than the Sun: Time to Abolish Nuclear Weapons by Scott Horton
Enough Already: Time to End the War on Terrorism by Scott Horton
Fool's Errand: Time to End the War in Afghanistan by Scott Horton
Diary of a Psychosis: How Public Health Disgraced Itself During Covid Mania by Thomas E. Woods, Jr.
Last Rights: The Death of American Liberty by James Bovard
Questioning the COVID Company Line: Critical Thinking in Hysterical Times by Laurie Calhoun
Domestic Imperialism: Nine Reasons I Left Progressivism by Keith Knight
Voluntaryist Handbook by Keith Knight
The Fake China Threat and Its Very Real Danger by Joseph Solis-Mullen
The Great Ron Paul: The Scott Horton Show Interviews 2004–2019
No Quarter: The Ravings of William Norman Grigg, edited by Tom Eddlem
Coming to Palestine by Sheldon Richman
What Social Animals Owe to Each Other by Sheldon Richman

Keep a look out for more great titles to be published in 2024.

www.ingramcontent.com/pod-product-compliance
Lightning Source LLC
Chambersburg PA
CBHW051621010526
44119CB00033B/438/J